Turnaround management

Tom Lenahan

OXFORD AUCKLAND BOSTON JOHANNESBURG MELBOURNE NEW DELHI

To my wife Liz
First, last and always
and to my sons
Julian and Craig
our hope for the future

Butterworth-Heinemann
Linacre House, Jordan Hill, Oxford OX2 8DP
225 Wildwood Avenue, Woburn, MA 01801-2041
A division of Reed Educational and Professional Publishing Ltd

A member of the Reed Elsevier plc group

First published 1999

© Tom Lenahan 1999

British Library Cataloguing in Publication Data
A catalogue record for this book is available from the British Library

Library of Congress Cataloguing in Publication Data
A catalogue record for this book is available from the Library of Congress

ISBN 0 7506 4283 1

FOR EVERY TITLE THAT WE PUBLISH, BUTTERWORTH-HEINEMANN
WILL PAY FOR BTCV TO PLANT AND CARE FOR A TREE.

Printed and bound in Great Britain by Biddles Ltd, Guildford and King's Lynn

Contents

Preface

Turnaround management is project management – it has all its main elements. It also has a number of features which make it unique.

Firstly, most other projects centre around the creation of something new, or the addition of something new to an existing entity, whereas turnaround management is concerned mainly with the replacement, repair or refurbishment of items which have malfunctioned in some way, or are worn, corroded or damaged.

Secondly, the work scope of other types of project is visible – in the sense that the project team know exactly what is to be done – whereas a large portion of turnaround work scope is hidden, the items to be worked being inside plant (vessels, machines etc.), covered by insulation or in inaccessible parts of the operating system.

Thirdly, in other projects any uncertainties are normally imposed by the operating environment (delivery of materials, availability of labour, the weather) but in a turnaround there is an additional (and sometimes highly significant) uncertainty which lies at the heart of the project itself, namely that the degree of wear or damage is unknown until the plant is opened up for inspection – and by then, if significant wear or damage is discovered, there is little or no time to remedy it without extending the duration of the project and driving up its cost. The best plan may therefore fail for reasons beyond anyone's control.

Finally there is the duration itself. Turnarounds are normally duration-driven, which requires the mobilization of hundreds or thousands of workers and a large quantity of materials and equipment on site, the completion of a substantial work scope (including work the need for which emerges during the event) and the demobilization of the men and equipment, all in a few short weeks. Due to the shortage (of time, safety, technical and logistics) problems are magnified and must be matched by a methodology which is equal to the task. The basic principle of that methodology is:

> *There are only two types of work on a turnaround, routine and unexpected. If the routine is under control there is time to deal with the unexpected but if the routine becomes unexpected the unexpected may become catastrophic.*

When I transferred from the construction to the chemical industry some twelve years ago to work as an overhaul manager, one of my first actions was to trawl the technical book catalogues to find the works which detailed the methodology for this complex business of plant shutdowns, overhauls and turnarounds. After all, I reasoned, there are thousands of plants around the world and each of them needs to be regularly overhauled. There should be such books; there was a need for them. To my surprise and dismay I could find none. Over the intervening years I occasionally repeated the trawl but always came up empty-handed. The company I worked for at that time (ICI) had a set of instructions detailing what had to be done to manage an overhaul but there was very little guidance on 'how to do it'. The individual engineers and managers did, however, have plenty of 'how to' material which they, like me, had developed over the years. So a lot of knowledge was there but it just didn't exist in a unified form. I suggested that we should pool all of the knowledge and create a coherent methodology. It took over a year to accomplish the task.

That was the beginning for me of the quest which culminated in this book. For over ten years I have expanded my knowledge by practice, by reflecting on that practice and by teaching turnaround methodology to others in various parts of the world. With some help from fellow toilers in the same field I have developed and simplified the material. I have never lost the belief that there should be a book dedicated to the management of turnarounds. Finally, here it is. I am not arrogant enough nor yet naive enough to believe that it is definitive, but it is a start, a foundation, if you like, upon which others can build. There is one significant difference between what I visualized all those years ago and what has come to pass – I never thought to see my name on the cover.

Tom Lenahan

Foreword

The planning and control of maintenance work has always been a major element of the courses which, over many years, I have taught in the field of maintenance management. I have always felt that if there were any particular merit in my approach then it probably lay (because of my own industrial experience) in its treatment of systems for the on-going planning and scheduling of the day to day routines – the servicing, the minor repairs and the smaller overhauls. I have, however, found it much more difficult to teach the organization of complete plant shutdowns, those that are done in order to carry out comprehensive programmes of preventive and corrective work.

Major shutdowns can involve hundreds, even thousands, of tradesmen, many diverse resources, planning horizons of many months and major cost penalties for duration over-runs. Although I was sure this subject should be part of our postgraduate programmes in maintenance management there was no relevant textbook that could be recommended, one which presented examples of real industrial practice. For many years my colleagues and I compensated for this by structuring the teaching of the subject around the presentation of critical path methods (and various other techniques) culled from textbooks on general project management.

It became increasingly clear to me that this important and complex topic – turnaround management – was one which deserved a textbook in its own right, a book which also could only be written by someone possessing many years' experience of the task, total familiarity with its basic procedures, and the ability to distil out its essential principles and concepts. Tom Lenahan more than meets this specification and has written a book which provides a major advance in the literature on this topic. It will be an indispensible guide for all who may be directly or indirectly involved in the organization of turnarounds, be they practitioners of engineering management or students of its complex arts.

I am pleased to be his friend and honoured to have had the opportunity to write this foreword.

Dr Anthony Kelly 1999

Acknowledgements

This book is the culmination of a ten-year quest to distil basic principles out of the rich and varied practice of turnaround management. Many people have, over the years, contributed concepts, insights and suggestions which have added to the sum of my own knowledge and served to clarify the central theme of this book, namely that turnaround management is a rational, coherent, process.

Special thanks first to three people: to Professor Anthony Kelly, Central Queensland University (and Honorary Fellow, Manchester University School of Engineering), who saw enough merit in my work to suggest that it was worth publication, to Simon Smith, Head of the Asset Management department of Eutech Engineering Solutions, who contributed time, materials and the introduction to Chapter 1, and to my son and colleague Julian, who organized the material for the initial draft of the manuscript and added some insights of his own.

Many thanks also to John Harris, Manchester University School of Engineering, who edited the complete text, to Michael Forster of Butterworth-Heinemann, and to the following colleagues for their part in the quest:

> Barry Stirling of Asean Bintulu Fertilisers
> Eric Scott of O'Hare International
> Chris Greaves of ICI Katalco
> Steve Waugh of ICI
> John Billington of Tioxide
> Ian Adams of Foster Wheeler
> Dr Marwan Koukash of EuroMaTech
> N. Sankara Narayanan of Hofincons India
> Claire Gulliver of ICI Paints

1
Turnaround overview: context and strategy

Introduction

The turnaround in a business context

It is perhaps an obvious statement to make that, given the complexity of the technology currently employed in the design and building of manufacturing plant, the limitations of the technology used to maintain it and the ever-increasing problem of its ageing, there will always be a need for maintenance work on the physical assets of the manufacturing activity. Furthermore, in an intensely competitive global market, characterized by increasing scales of production, the effective planning and management of that maintenance activity is coming to be seen as an ever more critical business process – one which is capable of differentiating the excellent performers from those who are merely capable, and from those who are less than capable.

It is also the case that with large scale assets there will be maintenance work which can only be carried out when the plant has been taken off line, decontaminated and made safe for the performance of such work (which includes, but is not limited to, critical inspection, equipment overhaul, repairs and plant modification).

The case for avoiding such interruptions to production, particularly in the case of continuous processing plants, is well made by those responsible for production and for the profitability of the company. As a consequence, a more serious and focused effort has recently gone into the design of technologies capable of monitoring and maintaining plants on line and hence minimizing costly outages with all their attendant risks to safety, reliability, business and the environment.

While the quest for production without regular plant shutdown goes on and the goal remains tantalizingly out of reach, there will remain the need – world wide and for the foreseeable future – to organize and carry out significant maintenance activity in the form of plant turnarounds on much of the existing industrial asset base. The purpose of this book is to explain, in a structured and logical manner, an approach to achieving this with minimum of risk to the enterprise.

The approach has been developed over many years by turnaround practitioners who have managed such events, large and small, on a wide variety of assets in Europe, the USA, Africa and Asia Pacific. These practitioners have generated, at both strategic and tactical levels, a corpus of principles, routines and processes that have proved to be a sound basis for managing these complex, hazardous and time-driven activities.

While it is clear that there have been major benefits in sectors such as oil and gas, petrochemicals, chemicals and utilities, the methodology will commend itself to other areas where the degree of sophistication may be less but where the business benefits may still be considerable, i.e. where large losses may be incurred by unnecessary planned downtimes.

It is interesting to conjecture why turnarounds have not hitherto received this level of attention, given their significant impact on many companies' performances around the world. The answer may lie in the fact that most companies have a history of tolerating higher than necessary downtime, with outages becoming even more frequent as their plants get older, and perhaps, even now, accepting an annual shutdown as a 'necessary evil'.

With so much revenue at stake it is probably not too surprising that it has been the process sector that has led the way out of this frame of mind by constantly striving to lengthen the intervals between turnarounds – from the traditional twelve or twenty four months to four, five and, in some cases, as much as eight years. This strategy has of course necessitated changes in plant operating standards and inspection techniques. In the UK this has been helped by the recent changes in the Pressurized Systems Regulations which have turned from the traditional prescriptive approach to something more flexible which, while allowing the companies which operate plants to use their technical expertise when setting intervals for inspection, nevertheless places the responsibility for the safety of the plant more squarely on the shoulders of those same operating companies. The corollary to this is that the planning and preparation for turnarounds has to be carried out ever more carefully, aligning capital programmes, scrutinizing and challenging work scope, assessing plant deterioration and its likely impact on reliability, planning stocks and safeguarding supplies to customers, and partnering with major engineering contractors specializing in plant overhaul (who are now generally the only available resource capable of carrying out the scale of activity required).

It is the potentially daunting nature of these activities and the long lead times required to manage them effectively that has given rise to this book. The following chapters therefore seek to explore firstly, the strategic issues surrounding the key questions – such as the fundamental need for turnarounds and how they will be managed – and secondly, the more tactical issues of how to plan, prepare and execute these events.

Reliability, the fundamental driver

In order to be profitable, a company needs consistent means of production delivered by reliable operating plant. There are various definitions of reliability and many company mission statements combine the words reliability, capability, utilization and the like in seeking to capture what it is they require from those who operate and maintain the plant. The general aim of this book is to strip techniques and processes down to their simplest form to lay bare what chemists used to call the 'active' ingredients – those which make the difference. In keeping with this principle the simple definition of reliable plant, for the purposes of this book, is

> *'a plant that is available when required and capable of performing to designed specification economically and safely'*

Again, in keeping with the purpose of this book, maintenance, a physical expression of asset management, is defined as

> *'the sum of activities performed to protect the reliability of the plant',*

activities which will therefore help provide a consistent means of production which will help generate profit.

Finally, a turnaround is defined as

> *'an engineering event during which new plant is installed, existing plant overhauled and redundant plant removed'*

Because the turnaround is a significant maintenance and engineering event a direct connection can be drawn between its successful accomplishment and the profitability of the company. There are a number of facets to this connection and a brief exploration of each will serve to set the turnaround activity in a business context.

Cost of the event

In this context, a turnaround impacts the business in a particular way. It is financed from company profits and, because it is typically an expensive event, it can have a major impact on those profits in the year in which it is performed. The cost of turnarounds performed at intervals greater than one year may be spread over a number of years to lessen the apparent impact on one year's profit, but the loss of profit remains the same. Determining the total cost of a turnaround is, at present, a subjective exercise because differences in culture from company to company mean that different factors and, in some cases, a different logic is used to determine it. For instance, should the profit lost because a plant is not producing for the period of the turnaround be considered as part of the turnaround cost? If not, why not? It is only when the total true cost of the event is known that its real impact on the business can be assessed.

Drain on resources

The turnaround is a drain on company resources and often diverts personnel from other important work. In most cases the event requires many more people than are normally employed on the plant and external resources have to be brought in.

Hazard to plant reliability

The turnaround is a potential hazard to plant reliability. Paradoxically, although it forms a major part of the maintenance strategy – the purpose of which is to protect the reliability of the plant – it can actually diminish or destroy reliability if not properly planned, prepared and executed, due to poor decisions by managers and engineers, bad workmanship, the use of incorrect materials and proprietary items, or because of damage done to the plant while it is being shut down, overhauled or re-started.

Potential safety hazard

The turnaround increases the potential for harm to people, property and the environment. Compare the normal routine on the plant to the situation which exists during the turnaround. Normal routine is characterized by the presence of a (relatively) small, experienced team performing familiar tasks using (it is to be hoped) well defined procedures in order to make products which will be sold at a profit. The turnaround reverses almost every characteristic of normal routine; the plant is shut down, taken apart and worked on by a large number of strangers using unfamiliar and inherently hazardous procedures and equipment. Under these circumstances the potential for accidents rises almost exponentially and must be matched by a safe system of working which minimizes the probability of loss. This adds to the cost of the turnaround and may even be dwarfed by the consequences to the company of actual loss, in terms of monetary cost or loss of reputation.

Risk of overspend and overrun

Finally, because of the technical uncertainty inherent in the turnaround due to the possible occurrence of unforeseen problems, there is an ever present risk of the cost estimate being exceeded, the duration of the event being extended, or both. The prudent company will therefore take this into consideration by building time and cost contingencies into the turnaround plan (which adds to the cost) – but even these may be exceeded. The resultant extra costs and loss of revenue may be substantial.

Strategy

Taking all of the above into consideration, as well as other factors which will be dealt with in this section, it is incumbent upon the senior management of the company to create a business strategy for managing turnarounds. The basic objective of this strategy – and this may seem an unusual perspective in a book dedicated to the management of such events – should be to eliminate maintenance turnarounds altogether. Thereafter, if it should prove that the event is absolutely necessary, senior management should ensure that the turnaround is aligned with maintenance objectives, production requirements and business goals.

Policy team concept

A company which is committed to getting the best value from its turnaround programme can realize this commitment in a concrete way. The most senior manager in the company should form a policy team (variously referred to as the *Steering Group* or the *Oversight Committee*) which should consist of senior managers who will take responsibility for the long-term strategy for turnarounds. The committee will meet at regular intervals throughout the year to review current performance and formulate high-level strategy for the management of turnaround events and the long-term turnaround programme.

Long-term view

On any plant which has been operating for a number of years some turnarounds will already have been performed, and in every plant covered by the imperatives of this book there will be future turnarounds. Rather than being seen as individual events separated by a number of months or years, turnarounds should be viewed by senior management as a sequence of linked activities (with many identical features) which are performed as part of the continuing process of asset management and are tied together, in the long term, by inclusion in senior management's business strategy.

Consideration of past practice, present intentions and future requirements gives a well-rounded perspective and prompts the types of strategic considerations which follow. Detailed questions for the following elements are contained in Figure 1.1.

Turnaround philosophy

To be truly aligned to overall business strategy, the evolution of asset management should be driven by the constant search for ways to change from the current vogue of preventive maintenance – which is driven by technical considerations – to a philosophy of maintenance prevention, which is driven by business needs. While it may not be possible to achieve zero maintenance, the only way to find out how nearly we can approach this utopian ideal is to question every maintenance practice to determine if it can be eliminated (by addressing the cause that generated the need for it). This being the case, it is logical to examine the largest maintenance initiatives, namely turnarounds, first.

The very first question (a negative answer to which would obviate the necessity for the succeeding questions) is

'is the turnaround necessary at all?'

This question – the fundamental one – should be given the serious consideration its potential payback deserves. It should not be dismissed lightly. It is a powerful and difficult question because it forces many managers to abandon the comfort zone they have inhabited for so long, with regard to the subject of maintenance in general and of turnarounds in particular.

Questions to be posed by senior management at an early date to minimise the impact of the turnaround on business performance

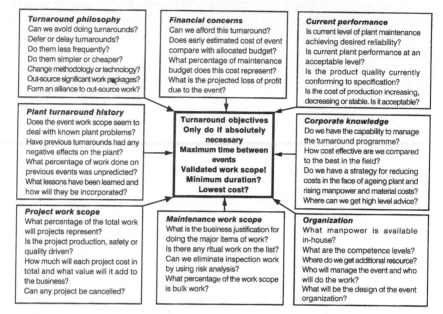

Turnaround philosophy
Can we avoid doing turnarounds?
Defer or delay turnarounds?
Do them less frequently?
Do them simpler or cheaper?
Change methodology or technology?
Out-source significant work packages?
Form an alliance to out-source work?

Financial concerns
Can we afford this turnaround?
Does early estimated cost of event compare with allocated budget?
What percentage of maintenance budget does this cost represent?
What is the projected loss of profit due to the event?

Current performance
Is current level of plant maintenance achieving desired reliability?
Is current plant performance at an acceptable level?
Is the product quality currently conforming to specification?
Is the cost of production increasing, decreasing or stable. Is it acceptable?

Plant turnaround history
Does the event work scope seem to deal with known plant problems?
Have previous turnarounds had any negative effects on the plant?
What percentage of work done on previous events was unpredicted?
What lessons have been learned and how will they be incorporated?

Turnaround objectives
Only do if absolutely necessary
Maximum time between events
Validated work scope!
Minimum duration?
Lowest cost?

Corporate knowledge
Do we have the capability to manage the turnaround programme?
How cost effective are we compared to the best in the field?
Do we have a strategy for reducing costs in the face of ageing plant and rising manpower and material costs?
Where can we get high level advice?

Project work scope
What percentage of the total work will projects represent?
Is the project production, safety or quality driven?
How much will each project cost in total and what value will it add to the business?
Can any project be cancelled?

Maintenance work scope
What is the business justification for doing the major items of work?
Is there any ritual work on the list?
Can we eliminate inspection work by using risk analysis?
What percentage of the work scope is bulk work?

Organization
What manpower is available in-house?
What are the competence levels?
Where do we get additional resource?
Who will manage the event and who will do the work?
What will be the design of the event organization?

Figure 1.1   Questions to be posed by senior management at an early date to minimise the impact of the turnaround on business performance

Every maintenance task carried out costs money that is subtracted from the bottom line, the profit margin. At the front end of the business, disciplines such as procurement, production planning, sales and marketing are being honed to a fine edge by use of the latest technology, massive management input and constant re-evaluation, while at the back end disciplines such as maintenance (and especially turnaround management) have undergone, in many companies, little more than cosmetic change in the past thirty years. A simple exercise to bring home the seriousness of this situation is to calculate the total cost (including the cost of plant-based personnel and the profit lost due to the plant being off line) of the last turnaround performed by the company and divide it by the estimated daily profit of the company. The result will indicate the number of days that the company had to work, on full production, simply to finance the activity. To use a football analogy, it's no use scoring goals at one end of the park if you are scoring own goals at the other end.

A common response to the above is the statement that, in the real world you cannot avoid the necessity to maintain the plant (turnarounds being, in part, simply large maintenance events). The crucial distinction in this statement is the definition of the term maintenance. It generates a number of questions, each of which is crucial to understanding the high cost of maintenance and turnarounds. They are:

1. How many of the tasks on a turnaround work list are generated by:
   - Legal requirements imposed upon the company?
   - Actual maintenance needs based on equipment manufacturers' recommendations and genuine requirements evolved by experience?
   - Fixing chronic problems which have never been investigated to find the root cause?
   - Ritual tasks which are simply unnecessary?
   - Avoidable attrition by process fluids and substances (corrosion, fouling etc.)?
   - Shortcomings inherited from design, procurement, installation and commissioning?
   - Inadequacies in the approved plant operation procedures?
   - Incorrect operation of the plant?
   - Insufficient environmental protection?
   - Lack of competence on the part of those (at any level) who operate and maintain the plant?
   - Insufficient preventive maintenance throughout the year?
   - Unrealistic expectations by the plant owners (overloading the plant)?

All such tasks generate cost, but are they all necessary? If they are accepted unquestioningly as 'maintenance' tasks then the definition of 'turnaround' must be expanded from the maintenance definition, i.e.

'the sum of activities performed to protect the reliability of the plant'

to

'the sum of activities performed to protect the reliability of the plant against all comers'

If, on the other hand, it is recognized that all of the tasks in the above list can be investigated with a view to eliminating, or at least minimizing, them, then the first step has been taken towards challenging the necessity for turnarounds. It should also be recognized that, in choosing to perform a turnaround (no matter how complex and difficult the planning and execution of the event may be for those at a lower level) the senior management may be sacrificing profit by choosing the easy option.

If, after the careful consideration outlined above, the senior management still believes it has no alternative but to perform the turnaround then the same challenging questions should be addressed at the level of the single task to ensure that the desired outcome is achieved with the minimum input.

Financial concerns

Turnarounds often cost more than is realized, because many companies do not calculate the total cost, concerning themselves only with the direct cost of planning and executing the event. Unless every element of cost thus generated is taken into consideration, control is not possible. We can only affect what we perceive.

The fundamental financial concern should be to plan and execute the event as cost effectively as possible. Using 'acceptable cost' as the tool with which to examine all expenditure will sharpen management focus and reduce the myriad elements of the turnaround to a single measurable and controllable unit – money.

If money is to be used as the measure then it is imperative that the scope of the turnaround work be effectively defined at the earliest possible date. The work scope is the foundation upon which all other aspects of the turnaround rest and will have a major influence in determining the final cost. The logic is simple; to exert control from the earliest possible date, senior management must have as accurate a cost estimate as possible – and this depends upon effective definition of the work scope.

An early cost estimate (even if only in 'ball park' terms) will indicate whether the budget for the event is sufficient. If not, senior management must choose between providing extra funds to cover the shortfall, eliminating elements of the work scope to bring the estimate back within budget, or finding some other strategy to optimize cost and work scope (the options are discussed in more detail in Chapter 6).

A second important consideration is how much of the overall maintenance budget should be allocated to the turnaround. This will depend upon the age and nature of the plant, the company's maintenance philosophy and the scope of the work. (A significant amount of benchmarking of maintenance costs has been done by companies such as Salomon Bros in the USA, and IPA in the UK.)

A third consideration is the amount of profit lost

(a)   while the plant is off line due to the turnaround and
(b)   on subsequent days when plant production rate or product quality may be below the required standard.

The final consideration here, a cost element often overlooked or hidden, is the wage and salary bill for plant personnel during the turnaround. It is argued by some managers that employees have to be paid whether the plant is operating or off line, and that the costs should therefore be allocated to production. The counter argument is that plant employees are paid to produce and if, during the turnaround, they are not producing then their cost should be allocated to the event, which otherwise will appear to cost less than it actually did (and, conversely, the cost of *production* will appear to be greater than it *actually* is).

Current performance

The scope of the turnaround work may be dictated by the current operating performance of the plant and the level and effectiveness of preventive maintenance carried out in the periods between turnarounds.

Rationalization and 'right-sizing' has led many companies to reduce their maintenance personnel to a minimum. One negative side-effect of this is that some preventive (and even some corrective) work does not get done. There is a great temptation to dump this into the turnaround work scope, to relieve the pressure on the maintenance crew, but it must be borne in mind that (due to higher management costs and loss of profit) it costs significantly more to perform a task during the turnaround than during normal operation, the work scope may be unnecessarily encumbered, resources may be diverted from more critical work and the duration of the event may be extended. The golden rule is:

> *The only tasks which should be allowed on to the turnaround work list are those which cannot be done at any other time.*

The turnaround should not be used to mask the shortcomings of the company's maintenance philosophy.

Regarding the work list which is generated by the current performance of the plant, care should be taken to ensure that the remedial work requested will actually address the problem being experienced.

Plant turnaround history

Turnarounds are performed to protect the reliability of the plant. The plant should perform as well or better after the event than it did before it. An examination of past turnaround performance (where this is possible) and of subsequent plant performance will indicate whether the turnarounds have provided the protection expected. If not, the situation must be reassessed to find out why and a new rationale for turnarounds developed. Past events should also be analysed to ascertain the ratio of emergent work (which only arises after the execution phase has begun) to planned work and the extent to which emergent work increased the planned expenditure (the norm is between 5 and 10 per cent, the worst case in the author's experience being in excess of 45 per cent). If past levels of emergent work are unacceptable, the senior management must ensure that plant personnel address the problem and improve the quality of technical specification in turnaround work requests.

Corporate knowledge

Turnarounds are a sequence of infrequent events in the life of the operating plant and they require management expertise significantly different from that required to operate and maintain the plant under normal circumstances. Ineffective management of the event can add more to its cost than any other factor.

Senior management must honestly assess whether their own people have the capability to manage and control the event. Even if technical capability exists in individuals, the cost effectiveness of internally managing the event must be analysed from the perspective of forming the technically competent individuals into a team capable of managing its total conduct. Again, a good start is to benchmark against the best in the field.

The expectation of senior management should be for the cost of each subsequent turnaround to be reduced – by the application of greater knowledge, superior technology and better organizational ability, even in the face of ageing plant and rising costs. This requires expert, professional, event management.

Project work scope

In terms of an operating plant, a project is a discrete package of work performed for the sole purpose of improving plant performance in some way (otherwise it is surely a waste of time, effort and money). Demands for projects are generated from various sources – statutory safety requirements, production or quality improvement programmes and the like. Senior management should evaluate each proposed project in terms of necessity, desirability and payback. The obvious questions are:

- Is the project work really necessary and vital to the safety, quality or production efficiency of the plant?
- Is it desirable at this time in the life of the plant; is it too late or too soon?

What will the payback be in terms of increased profit or reduced costs? The author once attended a de-brief meeting at which the business area manager stated that

> *'During the last few turnarounds the company has spent several million pounds on projects. The payback to date has been nil – the plant is not performing any better than it did before all this work was done'*

It is not sufficient for projects to be technically desirable, they must also be justified from a business point of view.

It is a rare turnaround that consists solely of maintenance tasks. More typically, the event will involve a mix of maintenance and project work. The ratio of one to the other will be determined by the current needs of the plant and the expectations of the management. The ratio is important because an event comprising 90 per cent maintenance and 10 per cent project would be managed differently from an event with the opposite proportions. Also an event involving a large number of small to medium sized projects would be managed differently from one involving only one very large project.

Whereas the turnaround manager is responsible for the planning and execution of all maintenance tasks, projects are planned and, in many cases, resourced by a project manager or engineer. This means that on a large event there could be a number of project teams who execute their work independently of the maintenance work. However, for the event to be coherent all of the work – maintenance and projects – should be integrated into a single plan. The task of senior management is to decide how this is to be done and how responsibility and authority are to be apportioned.

Maintenance work scope

Maintenance work can be divided into three approximate categories:

- *major tasks*, such as the overhaul of a large machine or the re-traying of a large distillation column, which require engineering input;
- *small tasks*, such as the cleaning and inspection of a small heat exchanger, requiring only the specification of the work that is to be done;
- *bulkwork*, the overhaul of a large number of small items such as valves and small pumps, which simply needs to be scheduled.

When evaluating the early work scope, the senior management should challenge the nomination of each major task to ensure there is a sound business justification for doing it in the particular turnaround or whether it can be eliminated altogether or deferred to a future event. Special consideration should be given to those tasks which will determine the critical path, and therefore the duration, of the event. This is the exercise of risk management and must be based on effective risk analysis.

Risk should be defined as:

> *the numerical probability of loss occurring.*

The two important questions to be asked are:

1. What is the specific nature of the harm (to people, production, property or the environment)?
2. What is an *acceptable* numerical probability?

There are a number of systems which may provide an objective answer to the latter question, by providing tables of probability and predicted failure rates, but in the end it comes down to this: what does the management of the company, at the time, believe to be an acceptable level of risk, given the set of unique circumstances which surround the operation of the plant? The answer depends upon the experience of the managers, the assumptions they use to frame their questions and the confidence they have in their own abilities.

In the context of the preceding paragraph, bulkwork may seem insignificant but the author would venture to suggest that almost every experienced turnaround manager has had at least one example of an event in which valuable programme time was lost, not because of any problem with any of the major tasks (which were probably planned down to the last detail) but because of the bulkwork which, due to its large numbers and wide dispersion, in some measure got out of control. The senior management should examine the strategy for scheduling and managing bulkwork and should assess its effectiveness.

Organization

The design of the turnaround organization will depend upon a number of questions concerning the planning, execution and management of the event which can only be answered by the senior management. There is a whole spectrum of types of organization which can be designed for the event (see Chapters 5 and 7). To illustrate this we will concern ourselves here only with the options which lie at either end of the spectrum. At one end is the option for the company to manage the event using their own managers and engineers. This begs the two questions:

- Are there enough people within the company to fill the required roles?
- Is the level of competence of each individual appropriate to the role required?

The second question may seem all too obvious but experience has shown that people are often assigned to jobs on the basis of their availability and with little regard to the question of their competence.

At the other end of the spectrum is the option to outsource the total turnaround package to a contractor whose business is the planning and execution of turnarounds. The main questions here are:

1. What is the contractor's track record – especially on financial claims against the client for variations to contract?
2. Can the management feel comfortable about putting their plant into the hands of a contractor without putting in place a control team to monitor performance? (A question which can become circular, because someone in the company is bound to ask whether – if the company has to field a team to monitor the contractor – it ought not use that team to manage the event directly and cut the contractor management out?)

Clearly the formation of the event organization requires much thought.

An engineering perspective

In the context of engineering, turnaround management may well be unique. Other functions – such as projects, construction and production – are concerned with creating something new, whether it be programmes, plants or products. Turnaround management, on the other hand, is mainly concerned with the replacement, repair or refurbishment of plant which is worn, damaged or malfunctioning in some way. Yet other functions can plan their work in the knowledge that if their plan *is* disturbed it will be by some external force. Turnaround planning, however, is by its very nature uncertain.

The planning and preparation which goes into a turnaround is done while the plant is on line. Faults may be hidden and may only emerge when the plant is shut down, opened up and inspected. This makes it difficult to predict the exact work scope, and so there is an inherent uncertainty regarding the costing, resource needs and duration of the event.

The psychology of turnarounds

Another issue that adds to the complexity of turnaround management is probably best termed the *psychology* of these events. Consider the differences, already discussed, between the normal routine of a plant and the situation which prevails during a turnaround from the point of view of the operators of the plant. Their plant which, as far as they are concerned, was running perfectly well, will be shut down (*no production – no profit*) and literally pulled apart (*a messy, confusing, situation*) by a large number of people who do not normally work on the plant (*strangers*) – this after a period of time when all of their requirements have been ruthlessly challenged (*by these strangers*) and their technical records and operating procedures will have been subjected to close, external, scrutiny (*and possibly found wanting*). On the other side of the fence there is the team charged with managing, planning and executing the turnaround. They are working hard in the plant team's best interest and are confused by the fact that the latter seem less than enthusiastic about their help, and often seem to be obstructing their best efforts.

The potential exists for friction between the teams. Management's task is to supervise the issue and ensure that all personnel are integrated into a single team – no mean feat. The integration can only be accomplished by briefing, training, negotiating, convincing and, above all, establishing good clean channels of communication. Failure to address these issues can result in the turnaround manager being faced with messy human problems which can be far more difficult to overcome than any technical obstacle.

Integration

The turnaround organization – the group of people who, under the overall direction of the turnaround manager will plan, prepare and execute the event – is formed by integrating five different types of knowledge and experience. These are:

1. *Local*: provided by plant personnel who are familiar with the technical records, operating procedures, history, past performance and current problems of the particular plant.
2. *Work management*: provided by engineers and planners who are familiar with all aspects of the initiation, preparation, execution and termination of turnarounds.
3. *Craft*: provided by (internal or external) maintenance personnel who are familiar with the tasks and activities required to overhaul plant items.

4.   *Specialist*: knowledge and experience in discrete technical tasks provided by the vendors' representatives and other specialist sub-contractors.

5.   *Management of discrete projects*: provided by experienced project managers or engineers.

The phases of a turnaround

A turnaround is normally considered to be an engineering event of relatively short duration, but it is only one segment of a cyclical process with four phases – initiation, preparation, execution and termination – each of which has its own specific set of critical issues and activities (see Figure 1.2). It is rightly referred to as a cycle because the initiation phase of the next turnaround should follow on from the termination phase of the current one.

Phase 1: Initiation

The reason for the inclusion of this phase (most other approaches start with the planning and preparation phase) is that this book attempts to consider the *entire* turnaround process from the moment some senior manager flags up the necessity to start considering the requirements for a forthcoming event (this could be two years or more before the event itself). It is therefore necessary to define, in detail, the strategic issues to be addressed and the activities required to move the process to the point where it can actually be planned and prepared. This phase is characterized by defining objectives, setting policy and appointing the necessary personnel to set up the preparation team and gather basic data.

Phase 2: Preparation

This is the major phase of the process. A small team of people work over a long period of time to specify, schedule, resource and cost the large volume of tasks required to perform the event. It contains an element of uncertainty because it involves some prediction of unknown conditions (of plant items) which, in turn, can involve everything from informed technical assessment to 'fingers crossed' guesswork. This is because the subject of the planning and preparation, the internal condition of the plant equipment, is hidden from view. One way of dealing with the uncertainty is to conduct an *analysis of contingencies*, i.e. to answer, in this case, the following questions:

- What faults are likely?
- How long will it take to put them right?
- How much will it cost?
- What impact will it have on duration?

A contingency analysis will provide only approximate answers, but they are, in many circumstances, the only guides available.

Critical issues	**Turnaround phases**	**Critical activities**

Critical issues

- Reviewing past events
- Setting objectives
- Formulating policies
- Minimizing costs
- Balancing constraints
- Delegating authority
- Monitoring performance
- Flexibility of approach

Phase 1 – Initiation

Period during which turnaround parameters are defined, core personnel appointed and basic data organized. Can be spread over a period of months

Critical activities

- Senior business manager forms a steering group
- Team appoints a turnaround manager as its agent
- Manager mobilizes and leads preparation team
- Preparation team collect and collate basic data
- Plant team issue work request forms and provide technical data, plant history and local knowledge
- Plant team issues the initial work list

Critical issues

- Leading a small team
- Translating policy into a defined project
- Defining contingencies
- Creating a plan
- Reviewing contracts
- Forecasting costs
- Setting targets
- Selecting people
- Formulating rules
- Delegating authority
- Monitoring performance
- Resolving issues

Phase 2 – Preparation

3 to 18 month period (depends on the size of the event) during which a large quantity of data, technical and non-technical, is validated and transformed into a set of plans that will be used to execute the turnaround

The vital element in this phase is the work list because it forms the foundation upon which all other elements are built i.e., safety, quality, costs, materials, equipment, resources, logistics and duration

Preparation is characterized by close attention to detail and accurate calculation

Critical activities

- Preparation team and plant team challenge and validate the work list
- Turnaround manager freezes the work list
- Planning officer and preparation team –
  – prepare job specification packages
  – identify and plan pre-shutdown work
  – identify and procure long delivery items then procure all other necessary items and materials
  – define contractor work and place contracts
  – organize site logistics and create plot plan
- Preparation engineer works up major tasks
- Turnaround manager optimizes the following –
  – work schedule and resource profile
  – organization chart
  – cost estimate and expenditure control system
  – contractor list and project list
  – safe system of work and quality plan
  – briefing document and site rules
- Turnaround manager submits plans to steering group for discussion, decision and approval
- Preparation and plant teams brief all personnel

Critical issues

- Co-ordinating actvities
- Controlling planned work
- Controlling emergent work
- Achieving duration
- Minimizing expenditure
- Monitoring safety
- Monitoring quality
- Resolving issues

Phase 3 – Execution

Normally 2 to 8 week period when planned work is carried out and monitored against the event schedule, duration, cost, quality and safety requirements

The emphasis during this time is on effective control of work

Critical activities

- Plant team shut plant down to a prearranged plan (product off) with support of execution team
- Execution team perform work to turnaround plan
- Control team define and cost emergent work
- Execution team complete all work
- Turnaround manager demobilizes execution team and mobilizes start-up team
- Plant team start up plant to a prearranged plan (product on) supported by the start-up team

Critical issues

- Analysing performance
- Reviewing work done
- Extracting learning
- Recommending changes
- Closing out report

Phase 4 – Termination

Normally 1 to 2 week period when the work is closed out and performance is reviewed

Critical activities

- Start up team clean the site and remove equipment
- Plant manager inspects plant and accepts handover
- Turnaround manager demobilizes start-up team
- Plant and turnaround manager organize debriefs
- Turnaround manager produces final report

Figure 1.2 Turnaround overview

The final act of preparation is to communicate the requirements of the turnaround to every single person who will be involved at any level. This is accomplished by a series of briefings carried out by the plant team and the turnaround team.

Phase 3: Execution

In this phase the planning and preparation are tested against reality. Was the planning accurate? Was the preparation adequate? Was the contingency analysis sufficient? Execution is characterized by the performance of a large volume of tasks by a large number of people of many skills and disciplines, in a limited space and at different levels simultaneously, under (sometimes severe) time and financial pressure. The effective control and co-ordination of work is of paramount importance.

Execution can be broken down into a number of sub-phases, as follows:

- shutting the plant down (removing inventory, decontaminating, cooling, isolating);
- opening the plant up (physical disconnection of items and removal of covers);
- inspecting the plant (visual and instrumental examination and report);
- installation of new items, overhaul of existing items, removal of redundant items;
- boxing the plant up (final inspection, replacement of covers and reconnection);
- plant testing (pressure tests, system tests, trip and alarm tests);
- starting the plant up (re-connecting services and re-introducing inventory);
- plant clean up and final inspection (removing all traces) of the turnaround.

In reality, some of these activities will overlap, but on any given task the activities will occur in this sequence and, more importantly, the transition from one activity to another involves, in most cases, the transfer of responsibility for the progress of the job from one person to another. This situation highlights a fundamental characteristic of properly co-ordinated events, namely *single point responsibility* which demands that the person responsible at any point in the process must fulfil three requirements, viz:

1. Check that the previous stage of the task has been properly completed before taking it over.
2. Ensure that the current portion of the task is carried out to specification.
3. Hand the task over to the next stage at the earliest possible time.

This ensures that, irrespective of the length or complexity of the task, there is always someone directly responsible for its quality and progress. It is the turnaround manager who, in the final analysis, is responsible for co-ordinating the total event and who must ensure that single point responsibility is understood and operated by his teams.

Phase 4: Termination

There are two separate elements involved in terminating the event. The first is ensuring that the plant is handed back in a fit condition and the second is the de-briefing of every member of the turnaround organization, the latter in order to capture the lessons to be learned from the event so that subsequent turnarounds may be performed more effectively.

These four phases go to make up the turnaround process and even this brief overview indicates the complexity of its management. To be effective, it must be initiated, prepared, executed and terminated using a rational methodology which will give repeatable results. The remaining chapters of this book detail such a methodology. Rather than simply describing the separate activities required to perform a turnaround they are organized in as strict a chronological order as possible, in order to give a sense of the way in which the process unfolds over a period of months.

2
Initiating the turnaround

Introduction

At some point in time the senior management of a company will make the decision to initiate the process which will eventually lead to a plant turnaround. With a large and complex plant, this may be up to two years before the event; for a small plant, as little as three months. Whatever the time scale, there needs to be a period when strategic decisions are made, key personnel appointed and basic data gathered to allow the planning and preparation phase to proceed in a coherent manner.

The policy team (see Figure 2.1)

A turnaround is a large, expensive and time-consuming undertaking. Even a small event is, in relation to normal operation, a large undertaking because, although the work scope may be small, all of the activities described here have to be performed. The event requires close planning and preparation

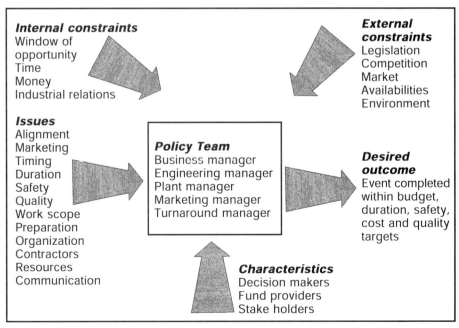

Figure 2.1 The policy team

and will, in all probability, involve every department in the company to some extent. It therefore requires serious consideration by the senior manager responsible for production to ensure that it is performed safely and cost effectively. Its complexity, however, puts it beyond the scope of any one individual and the senior manager therefore has to appoint a policy team (or steering group) the permanent members of which should be drawn from managers, executives and engineers who fulfil at least one of the following three criteria:

1. They are 'stake holders', i.e. are directly affected by the turnaround.
2. They provide the money to pay for the turnaround.
3. They have the authority to make decisions concerning the turnaround.

A typical policy team consists of representatives from the following functions (job titles have not been used because different companies use different titles for the same function or the same title for different functions). The main functions are:

- Senior management
- Marketing
- Production
- Maintenance
- Engineering
- Projects

The permanent team will also have the authority to call upon any other function to attend specific policy team meetings to provide particular expertise on an 'as-required' basis.

The purposes of the policy team (see Figure 2.2)

Once appointed, the policy team assume the responsibility of ensuring the success of the turnaround. As will now be explained, they have six specific purposes which focus their activities and decisions.

(1) The provision of funds for the turnaround

The team controls the budget and ultimately decides how much will be allocated to the turnaround. The actual cost estimate derived from the approved work scope will determine how much money is required. If the budget exceeds the estimate then the required amount may be allocated. If however, the reverse is the case then the team can pursue one of two options, either raise more capital to fund the approved work scope and duration or, by some means, reduce the work scope and/or duration to suit the budget. The team must also decide whether a contingency fund is to be set aside for unpredictable work which may emerge during the turnaround – and, if so, how much.

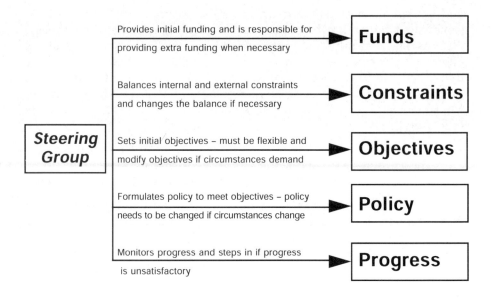

Figure 2.2 The purposes of the policy team

(2) The balancing of turnaround constraints

The turnaround will take place within a business environment which includes, but is not limited to, business performance, market circumstances, customers, competitors, the local community, available technology, legal requirements, health and safety at work, the local community and even the weather. The constraints are often in conflict with each other and have to be balanced. For instance, from a business point of view the cheapest and easiest way to get rid of surplus inventory while shutting a chemical plant down would be to dump it down a drain but that is (in most, if not in all, countries) illegal. Another option would be to flare it off – which, under certain circumstances, might not be illegal but the noise and flame of the burn might upset local inhabitants and adversely affect the company's standing in the community. The principle operating here is that every decision may achieve a benefit but will also incur a consequence. Benefits and consequences must be balanced in order to reach the optimum outcome.

(3) The setting of objectives for the turnaround

So that people know where to focus their efforts, any plan needs objectives. Turnaround objectives must be measurable and must deal specifically with quantity, quality, time, money and safety (QQTMS). These include, but are not limited to:

Quantity: How much work will be done or how many people will be employed?

Quality: What is the acceptable minimum level of workmanship?
Time: When will the event begin and how long will it take?
Money: What is the acceptable cost of the event?
Safety: What is the tolerable rate of accidents, incidents, emissions?

As with constraints, the first *four* objectives must be balanced, i.e. each objective must be set with reference to all the others and with due regard to current constraints. For instance, it is unrealistic to set a duration for the event of twenty-one days if the longest activity on the critical path is going to take twenty-three days; or if seven hundred men are required to achieve that duration and only five hundred are available.

The *fifth* objective, safety, is in a different class. The safety of people, property and environment must not be compromised to achieve other objectives.

(4) The formulation of policy to meet the objectives set

Policy is simply the course of action which is taken to achieve the objectives, and is applied to the same QQTMS elements. For example:

Quantity: A mixture of internal and external resources will be used because the former is insufficient to cover the event.

Quality: The ISO 9002 quality standard procedures will be adopted, including audits of all sub-contractors to ensure uniformity.

Time: Because of market demands, the window of opportunity is limited to twenty-two days (between 4th and 26th of September).

Money: A 40 per cent increase on the allocated budget will be borrowed from an external source to meet the increased work scope.

Safety: A professional safety manager will be engaged for the event because the company cannot field anyone with the appropriate experience.

The policy decisions taken at this level and at this time will affect every aspect of the turnaround throughout its four phases.

(5) The monitoring of progress against objectives

As mentioned before, objectives must be measurable so that progress can be monitored and compared against some standard to ensure that all phases of the turnaround are kept on track. Again, progress is measured against QQTMS in terms of quantities fulfilled to date, compliance with quality requirements, key dates achieved, current cost expenditure and daily accident/incident rates. The policy team must formulate a set of procedures to accomplish this (see example later in this section).

Where a written set of procedures exists, formal audits may be used as a monitoring device. Typically, a large event might have a sequence of six audits as will now be described. (Smaller events should be subjected to at least one preparation, one execution and one termination audit, while very large ones might benefit from several extra audits, the guiding principle being to perform the minimum number of audits required to ensure that preparation, execution and termination of the event are adequately performed).

Audit 1 Strategy audit

Carried out at the earliest possible date. It examines the strategy formulated by the senior management and evaluates it against the stated business aims of the turnaround.

Audit 2 System audit

Examines whether the systems that have already been put in place – or are being considered for later inclusion in the programme – will be adequate for effective implementation of the policy team's strategy.

Audit 3 Compliance audit

Examines planning and preparation, in all their aspects. Assesses whether they have been adequate to achieve the objectives of the turnaround and, in particular, whether the personnel involved are complying with the systems.

Audit 4 Preparedness audit

Carried out one to two weeks before the start of the event, it examines the state of readiness of the turnaround organization and the systems that will be used during execution to retain control of the event.

Audit 5 Execution audit

As its name implies, is carried out during the execution phase and examines the control system to assess whether the turnaround manager and his organization are in control of the event. It also examines the actual work methods being employed and evaluates them for compliance against the planned procedures.

Audit 6 Final audit

Carried out during the termination phase. Examines the effectiveness of the de-briefing system in capturing the lessons to be learned and also of the system for ensuring that the information is recorded in such a way that it will be available for future events.

It is up to the policy team to decide if they wish none, some or all of the audits to be carried out and – for those that are to be performed – whether they should be carried out by independent auditors or by company personnel. The timing of the preparation phase audits is dependent upon the length of the preparation period.

(6) The modification, if necessary, of objectives or policy

Because a turnaround is a complex process performed over a long period, the policy team needs to be flexible in its approach, able to modify its objectives or change its policies so that it can readily deal with changes (which may be radical) in process or environmental constraints. For example, the cost of borrowing the money to increase the budget by 40 per cent may suddenly be subject to a sharp rise in interest rate. Or new safety legislation may force an increase in the duration of the event.

The policy team needs to deal with these disturbances as they occur. The most acute disturbance is one that occurs *during* the actual event. A fault may be found in a critical piece of equipment and there may be no option but to carry out an extensive repair which will be costly and will increase the duration of the event. The problem here is one of *force majeure* because there is nothing that can be done to avoid it. In addition, the team may have very little time to react. The ability of the team to act competently will determine the success or failure of every other aspect of the turnaround process, hence the strict criteria for the selection of personnel to serve on the team.

The methodology

To ensure the purposes of the policy team are met, the team should adopt a rational methodology comprising the following basic elements.

(1) Creation of the policy team – at the appropriate time

Depending upon its size and complexity the turnaround may require a preparation period of anywhere from 3–18 months, and the policy team should be set up before the start of the preparation phase. Apart from the amount of planning and preparation required, this time scale is also determined by such things as the need to purchase expensive proprietary items which are on a long delivery (e.g. in some parts of the world the delivery time for a compressor rotor, say, may be as much as 16 months), the requirement for pre-shutdown work that has to be completed well in advance (e.g. the excavation and installation of foundations with a long cure-time) and the design and procurement of pre-fabrications. If the team do not have enough time to make their decisions the turnaround may be adversely affected, sometimes catastrophically so.

(2) Appointment of a full time turnaround manager

The complexity of the turnaround process and the fact that the permanent members can only operate on a part-time basis (because they still have their normal jobs to do) make it imperative that the team appoint a full time turnaround manager who will act as the policy team's agent in all matters. The team invest authority in him to act on their behalf and to make decisions – within the confines of the policy as laid down by them. Once appointed, he should take over as chairman of the team and set the agenda for each meeting.

The turnaround manager becomes the 'active' member of the team and ensures that issues are resolved by the *Proposal–Discussion–Decision–*

Action–Feedback routine. That is, he proposes a course of action; the team discuss it; they take a joint decision on the course of action; he ensures that the action is carried out and then reports the results of the action back to the team (a detailed description of the turnaround manager's role is given later in this chapter).

(3) Creation of a formal agenda

In order for the steering group to address the myriad aspects of the turnaround they require a formal agenda which covers all critical points. Typically, this would include (but not be limited to) the following:

1.   CONSTRAINTS
1.1   Review of previous actions (including review of previous events)
1.2   Window of opportunity and timing constraints
1.3   Duration
1.4   Budget and cost estimates (Are there differences? How can they be resolved?)
1.5   Comparison of workloads (Is this event similar to previous events – or not?)
1.6   Availability of resources
1.7   Special constraints

2   OBJECTIVES
     Quantity–Quality–Time–Money–Safety

3   WORK SCOPE
3.1   Legal requirements – if any
3.2   Major tasks
3.3   Preventive tasks
3.4   Corrective tasks
3.5   Condition monitoring
3.6   Modifications
3.7   Projects
3.8   Demolition

4   PREPARATION
4.1   Preparation plan
4.2   Work specification
4.3   Material procurement
4.4   Contractor selection
4.5   Integrated turnaround plan
4.6   Turnaround organization
4.7   Site logistics
4.8   Cost profile
4.9   Safety

A formal agenda ensures business is carried out in a rational manner and reduces uncertainty.

(4) The holding of regular team meetings

The best way to gain commitment from team members and drive the progress of the turnaround is to hold regular formal meetings, the frequency of which will depend upon the preparation time available and the size and complexity of the event. For a large event the team might meet weekly for the first four weeks to clear off the bulk of early work. Then monthly until one month before the event, when it would revert to weekly.

(5) The minuting of every action and decision

Every meeting must be formally minuted, special attention being given to two particular elements, as below.

(i)   When any action is delegated it should be recorded in full together with the name of the person who will carry it out and the required completion date. If the person responsible does not attend the team meetings, the turnaround manager must undertake to inform that person, in writing, of the requirements of the team. At each meeting the first point on the agenda must be a review of previous actions to ensure that each one has either been completed or is progressing satisfactorily; if not, the reasons must be discussed and decisions taken as to how best to resolve the impasse.

(ii)   Once decisions are agreed upon they should be written down formally, and every member of the team should be committed to them even if the agreement is not unanimous. The time for argument is at the discussion stage, not after a decision has been made.

There is a favourite saying among auditors 'if it wasn't written down it didn't happen'.

(6) *Formulation of a preparation plan* (see Figures 2.3, 2.4(a) and 2.4(b))

Because the planning and preparation of the turnaround are carried out over a long period of time, and include a large number of critical activities, the team requires a preparation plan, to monitor progress on a regular basis. Figure 2.3 shows the key events in the initiation and preparation phases, including both *strategic* events, that are the responsibility of the senior management and policy team, and *tactical* events which are the responsibility of the turnaround manager. Figure 2.4(a) is in the form of a Gantt chart where the timings and durations of turnaround team activities are displayed. Figure 2.4(b) is a network plan display with activities on the lines and due dates written in the nodes. What all three diagrams have in common is that they all define what has to be done, when it has to be done by and the dependencies between activities. Over the period of preparation the policy team, and more especially the turnaround manager, will use these documents to monitor progress and take any necessary action.

(7) *The reviewing of progress on all issues*

With so many issues at stake there needs to be a constant review of every aspect of the turnaround process. Every policy team meeting should involve a review process for ensuring that activities completed to date are consistent with the current strategy.

The turnaround manager

The turnaround manager is the key role in the organization. It is his responsibility to ensure all activities are carried out in the initiation, preparation, execution and termination of the event. Depending upon the company's situation he may be:

- *A company employee* – where there is a member of staff with the relevant knowledge and experience.
- *A consultant manager* – if the company does not have anyone with the requisite knowledge and experience (several consultancies provide this service).
- *A manager from the main contractor* – if a contractor is being used to perform the work they can normally provide a suitable candidate.

Personality

If the endeavour is going to be successful the person selected for the role of turnaround manager should possess the following character traits:

- Leadership, team building and negotiating skills;

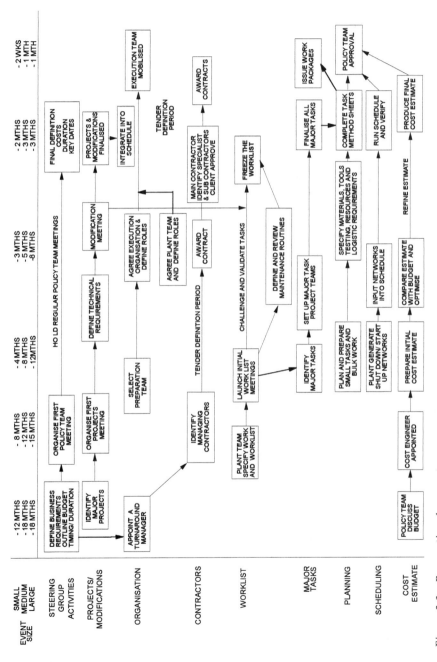

Figure 2.3   Preparation: key event programme

			Issue No.................
			LEGEND
Project name		Start Date	Duration ▬▬▬
			Float 0000000
Compiled by	Signature	Date	MUST date ▽
Approved by	Signature	Date	Dependency ....

Week: 01 02 03 04 05 06 07 08 09 10 11 12 13 14 15 16 17 18 19 20 21 22 23 24 25 26 27 28

1 DEFINE WORKSCOPE & CONSTRAINTS
 1.1 Collate work lists and projects.........
 1.2 Validate & delete unnecessary work.........
 1.3 Obtain approval and close worklist ....

2 PREPARE JOB SPECS (IF REQUIRED)
 2.1 Select Turnaround planning team.........
 2.2 Plan major tasks & bulkwork specifications.........
 2.3 Issue for comment, amend & finalise.........

3 IDENTIFY PRE SHUTDOWN WORK
 3.1 Order long delivery items & fabrications.........
 3.2 Order pre shutdown services.........
 3.3 Expedite, receive, store & protect 3.1 & 3.2.........

4 IDENTIFY & PROCURE MATERIAL & ITEMS
 4.1 Define material & item requirements.........
 4.2 Select suppliers & place orders.........
 4.3 Expedite, receive, store & protect.........

5 DEFINE CONTRACT WORK PACKAGES
 5.1 Split work into manageable packages.........
 5.2 Define types of contractors.........
 5.3 Create discrete contractor workscopes.........

6 SELECT TURNAROUND CONTRACTORS
 6.1 Prepare & issue ITBs to contractors.........
 6.2 Clarify & negotiate contracts.........
 6.3 Select contractors & let contracts.........

7 CREATE INTEGRATED TURNAROUND PLAN
 7.1 Receive & collate contractor work plans.........
 7.2 Create, run, refine & re-run schedule.........
 7.3 Issue for comments, amend and finalise.........

8 DEFINE TURNAROUND ORGANISATION
 8.1 Define "black box" organisation chart.........
 8.2 Create initial detailed organisation chart.........
 8.3 Issue for comment, refine & finalise.........

9 CREATE SITE LOGISTICS PROGRAMME
 9.1 Identify all site support requirements.........
 9.2 Collate requirements into a rational plan.........
 9.3 Order, receive & place requirements.........

10 FORMULATE TURNAROUND COST PROFILE
 10.1 Formulate initial turnaround costs (±20%).........
 10.2 Create detailed cost profile.........
 10.3 Issue for comment, refine & finalise.........

11 ESTABLISH PERMIT TO WORK SYSTEM
 11.1 Define permit to work philosophy.........
 11.2 Select permit to work issue team.........
 11.3 Block out permits pre shutdown.........

12 DEFINE SAFETY PROGRAMME
 12.1 Define turnaround safety philosophy.........
 12.2 Select team & define safe system of work.........
 12.3 Set up safety organisation on site.........

13 DEFINE QUALITY PROGRAMME
 13.3 Define turnaround quality philosophy.........
 13.2 Select team & define methodology.........
 13.3 Set up quality team on site.........

14 BRIEF TURNAROUND ORGANISATION
 14.1 Collate individual briefing documents.........
 14.2 Produce an integrated briefing package.........
 14.3 Brief everyone in the organisation.........

Figure 2.4(a) Turnaround preparation programme

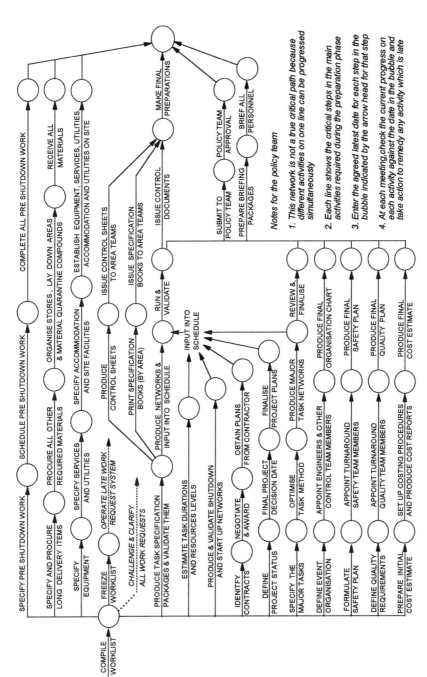

Figure 2.4(b) Turnaround preparation network

- A good working knowledge of turnaround type projects;
- Planning and co-ordinating ability;
- A flexible approach to complex problems;
- The ability to work effectively under pressure;
- A strongly developed sense of humour.

It takes patience and tenacity to manage the hundreds (sometimes thousands) of different individuals involved in the many difficult, and occasionally novel, situations that arise during the four phases of the turnaround.

Roles and responsibilities

The specific roles and responsibilities of the turnaround manager alter as the project advances through its different phases. They include (but are not limited to):

INITIAL PHASE
- Convening the initial meeting of the working policy team and chairing all subsequent meetings;
- Advising the policy team on specific turnaround requirements;
- Ensuring the collection and collation of basic data;
- Selecting and managing the turnaround preparation team.

PREPARATION PHASE
- Convening all necessary meetings;
- Monitoring preparation, progress and information feedback to the policy team;
- Validating work scope and freezing the work list;
- Approving the selection of contractors and vendors;
- Organizing, optimizing, finalizing and presenting the following plans to the policy team for discussion, approval and action:
  — Turnaround schedule
  — Organization chart
  — Safety plan
  — Quality plan
  — Cost estimate
  — Communication and briefing package
- Organizing the turnaround briefing programme.

EXECUTION PHASE
- Organizing assistance for the plant team during plant shut down;
- Managing the daily conduct of the turnaround;
- Taking action to ensure targets and objectives are met;
- Delegating specific responsibilities, tasks and activities;
- Chairing the daily control meetings;
- Reporting significant issues to the policy group;

- Setting priorities and trouble-shooting;
- Negotiating changes in resources and work scope;
- Organizing assistance for the plant team during plant start up.

TERMINATION PHASE
- Demobilizing resources and equipment, and organizing the cleaning of the site;
- Co-chairing debrief sessions with the plant manager;
- Collating all information and writing the turnaround final report.

Throughout the changing phases of the lengthy event, the one constant key figure is the turnaround manager.

The preparation team: membership (see Figure 2.5)

Purpose

During the initial phase of the turnaround its manager should appoint a preparation team who will be responsible to him for the detailed planning and preparation of all aspects of the project. It will consist of a small number of permanent members to carry out the general work, backed up by temporary members co-opted to perform specific functions. The size of the team will depend upon the size and complexity of the turnaround. For present purposes it is assumed that we are dealing with a large event requiring a full strength team. For smaller events the team would be scaled down, with some of the roles combined for performance by one person, to suit the available funds and the lesser amount of work. In addition to the turnaround manager, the permanent team would encompass those roles and responsibilities which will now be briefly summarized (but will be discussed in more detail later).

Note: The nomenclature used here is one which has been arbitrarily adopted by the author, there being no universally agreed one. It is the function which is important, not its name.

Preparation engineer

Whereas the turnaround manager has many duties which require him to be mobile, the preparation engineer would be permanently located on plant with the team. Typical roles and responsibilities would include, but not be limited to:

- Reporting directly to the turnaround manager;
- Liaising with plant and other necessary personnel on a daily basis;
- Managing the daily conduct and performance of the team;
- Organizing and chairing all preparation team meetings;
- Allocating work packages to team members;
- Validating the work list and referring it to the turnaround manager for approval;

Roles and responsibilities

1. Turnaround manager
- selects the members of the preparation team
- ensures the team is properly resourced, equipped, safety inducted and introduced to plant personnel
- sets out the team's targets for preparation
- attends nominated preparation meetings
- resolves strategic issues for the prep engineer
- delegates authority to the prep engineer
- monitors team performance via the prep engineer
- reports directly to the turnaround policy team
- refers policy issues to the policy team and assists the team to resolve them

2. Preparation (prep) engineer
- manages the daily conduct of the team
- chairs all preparation team meetings
- liaises with the client on a daily basis
- allocates work to team members
- validates the worklist and refers it to the turnaround manager for final approval
- plans all major tasks
- resolves operational issues
- monitors the team's performance and adjusts what is required to stay on target
- reports directly to the turnaround manager
- refers strategic issues to the turnaround manager

3. Planning officer
- assists preparation engineer to validate work list
- liaises with client planning organization
- allocates planning tasks
  supervises the work of the planners
- checks completed task sheets and forwards them to the plant manager for validation
- inputs networks and data into the scheduler
- creates and optimizes the turnaround plan
- assists the prep engineer planning major tasks
- resolves all planning issues
- reports directly to the preparation engineer
- refers operational issues to the prep engineer

4. Planning team
- gathers basic data for planning from the plant
- plans and networks small tasks
- produces task sheets
- specifies bulk work and lists on control sheets
- collates all documentation for task packages
- produces material requirement sheets
- produces equipment requirement sheets
- liaises with site logistics team
- prepares area task books for area co-ordinators and supervisors
- reports directly to the planning officer
- refers planning issues to the planning officer

5. Site logistics officer
- liaises with plant personnel to gather basic data on the plant and available surrounding land
- draws up a master plot plan showing locations of all turnaround logistic requirements
- produces various versions of the site plot plan for use by different disciplines
- liaises with everyone involved in the turnaround
- arranges all outdoor lay down areas
- sets up turnaround stores and formulates receipt and issue procedures for materials and equipment
- procures/receives, locates, protects and maintains:
  — materials and proprietary items
  — tools and technical equipment
  — transportation and cranage
  — services and utilities
  — accommodation and facilities
- controls hazardous substances
- provides for the daily needs of all personnel
- allocates tasks to the site logistics team
- supervises the daily work of the team
- monitors team performance
- reports directly to the preparation engineer
- refers operational issues to the prep engineer

6. Site logistics team
Usually semiskilled personnel –
- reports to the site logistics officer
- carries out all tasks set by the site logistics officer

Figure 2.5 The preparation team

- Planning all major tasks;
- Resolving day-to-day operational issues on behalf of the team;
- Monitoring team performance and adjusting requirements where necessary;
- Referring strategic issues to the turnaround manager for resolution.

Planning officer

The planning officer works closely with the preparation engineer. Typical roles and responsibilities would include, but not be limited to:

- Reporting directly to the preparation engineer;
- Liaising with plant planners and other necessary personnel;
- Assisting the preparation engineer with validation of the work list;
- Allocating planning tasks;
- Supervising the work of the planners;
- Checking completed planning documents and passing them to plant personnel for validation;
- Checking critical path networks produced by the planners and putting them into the schedule;
- Creating and optimizing the turnaround schedule;
- Assisting the preparation engineer to plan major tasks;
- Resolving all planning issues on behalf of the team;
- Referring operational issues to the preparation engineer for resolution.

Planning team

The number of planners is dictated by the amount of work to be done and the time and money available. Typical roles and responsibilities would include, but not be limited to:

- Reporting directly to the planning officer;
- Gathering basic data for planning from plant and other necessary personnel;
- Producing task specifications and networks for all small tasks;
- Specifying bulk work tasks and listing them on control sheets;
- Producing material requirement sheets for specified tasks;
- Producing equipment and services requirement sheets;
- Liaising with site logistics team;
- Preparing task books for execution team superintendents and supervisors;
- Referring planning issues to the planning officer for resolution.

Site logistics officer

Site logistics is concerned with the reception, location, protection, distribution and final disposal of all items and services required for the execution of the turnaround. Typically, the site logistics officer's roles and responsibilities would include, but not be limited to:

- Reporting directly to the turnaround manager;
- Defining the plant boundaries and the available surrounding land;
- Drawing up a site plot plan showing all logistics requirements;
- Setting up stores, lay-down and quarantine areas;
- Receiving on site, locating, protecting, distributing and disposing of:
    - materials, consumables and proprietary items
    - tools and equipment
    - vehicles and cranage
    - services and utilities
    - accommodation and facilities
- Controlling hazardous substances;
- Providing for the daily needs of all personnel;
- Allocating work to the site logistics team;
- Supervising the daily work of the team;
- Monitoring team performance;
- Resolving logistics issues;
- Referring operational issues to the preparation engineer for resolution.

Site logistics team

This team is normally made up of semiskilled and non-technical personnel such as storemen, drivers, cleaners etc. They report to the logistics officer and carry out the tasks set by him.

Preparation team: gathering basic data

Once the preparation team has been appointed they are normally accommodated in or near the plant offices because they will spend a great deal of their time on the plant. Manufacturing plants are hazardous places and it is likely that some or all of the team will be unfamiliar with the plant. It is therefore the responsibility of the turnaround manager to arrange the following safety precautions for any team members who are not existing members of the plant team.

Plant safety induction

The safety induction should familiarize the team with the hazards presented by the plant and with the operation of its emergency procedures (include the various sounds of the fire, toxic release and evacuation sirens). It must give adequate information regarding the permit-to-work system because members of the team may have to visit parts of the site controlled by permit in order to assess a particular job.

Plant tour

Experienced personnel should conduct the team on a tour of the plant, to familiarize the members with its geography and its hazardous areas.

Temporary site pass

This will last for the duration of the team's stay and will ensure that they are formally recorded as working on the plant.

Provision of safety gear

To ensure the team can be dressed to withstand the normal hazards of the plant they must be equipped both with normal safety gear (coveralls, hard hat, steel-toe-capped boots and light eye protection) and any plant-specific 'special' safety gear (which may include goggles, gloves, respirators etc.).

Data gathering

A turnaround is a complex process and like any other the quality of its outputs depends upon the quality of its inputs. The inputs and outputs of the preparation team are information. Data gathered from plant personnel and records are transformed into the specifications and plans required to carry out the turnaround. Unfortunately, on many plants, especially older ones where there have been several turnovers in personnel – and many modifications to equipment, procedures, systems and processes – some necessary data may either not exist or, if it does, may be out of date or unreliable in some other way. It is therefore essential that the basic data gathered from the plant and other sources is checked and validated by the team members before being used. There are two basic ways to gather data from people: by personal contact and through formal meetings.

Personal contacts (see Figure 2.6)

In order to gather accurate data, a network of personal contacts should be built up. The personnel involved would typically include – but not be limited to – Plant Manager, Safety Officer, Engineering Manager, Maintenance Manager, Project Manager, Chief Inspector, Stores Controller, Process Supervisors, Maintenance Supervisors, Electrical and Instrumentation Engineers and Supervisors. Rigging Supervisor, Scaffolding Supervisor, Civil Supervisors and the Plant Planning Team Leader. These are the people who possess the information the preparation team requires; building solid working relationships with them is important. Another issue to consider is that on any plant there will be a small number of individuals (sometimes just one person) who have worked on it for a long time and understand many of the ways that it works and why it sometimes doesn't. They will know the history of the plant and will be aware of its chronic problems. They will know where records are and will be able to advise the team on the validity of data. It is vital that they be sought out, recognized, cultivated and their knowledge and experience exploited to the benefit of the team.

Meetings to be kicked off include, but are not limited to:

Comments

1. Policy team meeting	
2. Plant standards meeting	
3. General worklist meeting	
4. Major task review meeting	
5. Project review meeting	
6. Inspection review meeting	
7. Spares review meeting	
8. Shutdown/start-up meeting	
9. Safety review meeting	
10. Quality review meeting	
11. Site logistics meeting	

Figure 2.6 Basic data gathering

Formal meetings (see Figure 2.7)

The very mention of the word 'meeting' tends to send a chill down the spine of most engineers. Meetings are often seen as wasteful, time consuming and ineffective. Turnaround preparation requires many formal meetings because they remain the best (and may be the only) way for a group of people to transmit, check, challenge and validate information.

There are a number of principles which guide and enhance the effectiveness of preparation meetings.

First –   the agenda of each meeting should be confined to a single subject so that attendees can focus their minds on a particular requirement.

Second – only those who are directly involved with that particular aspect of preparation should be invited to attend.

Third – each meeting should have either a clearly defined block of work to accomplish, or a defined time limit, or both.

Fourth – the meetings should all be chaired by the appropriate member of the turnaround preparation team, who then sets the agenda and takes the minutes.

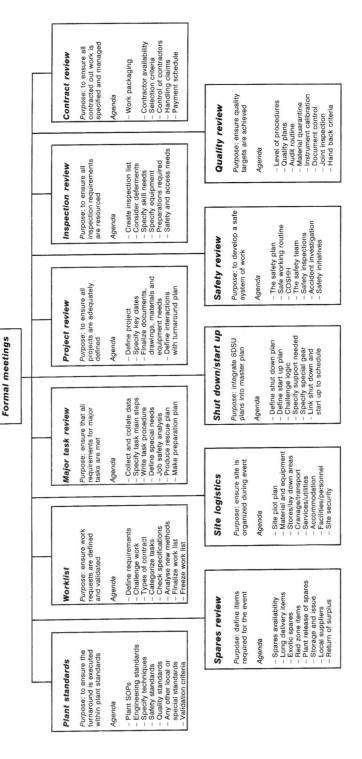

Figure 2.7 Formal meetings

Data required includes, but is not limited to:

Comments

1. Plant organization chart	
2. Plot plan of the plant	
3. Previous turnaround reports	
4. Previous budget/cost report	
5. Plant permit to work system	
6. Statutory inspection list	
7. Mechanical work list	
8. Electrical work list	
9. Instrument work list	
10. Trip and alarm schedule	
11. Civil work list	
12. Machine tasks	
13. Equipment cleaning tasks	
14. Projects and modifications	
15. Shut down/start-up network	
16. Access to plant P and IDs	
17. Access to line diagrams	
18. Access to plant history	
19. Access to technical library	
20. Access to spares lists	

Figure 2.8 Documentation

Some subjects (e.g. plant standards) may require only one meeting, while others, involving the resolution of complex issues (e.g. work listing) might require the convening of many.

Typically, the meetings involved in turnaround preparation would include, but not be limited to, the following subjects: turnaround policy; plant standards; work list; major task review; project review; inspection review; spares review; safety review; quality review and site logistics. In addition to these there will be a large number of *ad hoc* meetings, convened to resolve specific issues (e.g. the resolution of single design, manufacturing or delivery problems).

Documentation (see Figure 2.8)

As with meetings, there are a large number of documents involved in preparing a turnaround. These may be either in hard copy or on a computer data base. They will include, but not be limited to, plant organization charts, piping and instrumentation diagrams, technical library, spares lists, previous turnaround technical and cost reports, permit to work procedures, statutory inspection list, mechanical/instrument/electrical/civil work lists, machines task list, equipment cleaning tasks, trip and alarm schedule, shutdown and start up networks. The team needs to gain access to all of this documentation.

Conclusion

The initiation phase is the period during which the groundwork is laid for the preparation phase. The preparation team is selected, re-located to the plant site and inducted. The team members then make contact with key people on and off the plant, organize the initial meetings for all the various aspects of the turnaround and begin to acquire the necessary documentation and other information which will allow them to plan and prepare for the event. It is, as it were, the foundation of the turnaround process.

Case study

The total responsibility for the turnaround on chemical plant A – one of a group of five owned by the same company – fell on the maintenance manager. The senior management's involvement was to allocate a budget based on what was spent on the last turnaround, three years previously, plus a sum to cover for inflation and to appoint the maintenance manager as turnaround manager. Thereafter he was left to get on with it.

The work list was formulated by plant personnel, and the job of the turnaround planning team (who had been seconded from the other four plants in the group) was seen as simply planning and scheduling the work. There was little contact between the maintenance manager and the plant manager (who considered maintenance as a necessary evil), both of whom had been in their jobs for less than two years. The turnaround plan was prepared and a cost estimate generated one week before the start of the event. To the turnaround manager's horror the estimate was 64 per cent above the budget allocation and the planned duration was three days' longer than the time slot allocated by the business manager for the event.

Three days before the event was due to start an emergency meeting was finally called. It was attended by the plant manager, the maintenance manager and the divisional director for production. The last of these demanded to know why the cost of the turnaround had risen so sharply. The maintenance manager explained that there were several large packages of work which were only carried out every second turnaround so they had not been included in the previous one (from which the budget had been formulated). He had only found this out when he had investigated the reason for the sharp increase in cost. No one had informed him of the significant difference in work scope between the previous and present events.

The director asked if the extra work could be eliminated from the event. The maintenance manager replied that he just did not know at this stage and there was not enough time to do a worthwhile assessment.

The director then asked the plant manager if he could contribute anything useful. He replied that he also could not say whether the work needed doing or not – in any case it was a maintenance issue not a production one. He did offer the advice that if someone in the past felt it was necessary to do it then it probably was needed, and if it was not done the consequences might be serious. The director resigned himself to allowing the work to be done and left the meeting after roundly castigating both managers for their incompetence – he was not looking forward to the meeting with the board at which he would have to inform them that there was going to be a serious budget and duration overrun. ...A month later, the story was worse. Because a large amount of emergent work had arisen the final bill was 97 per cent greater than budget and there was an over-run of five days.

In the light of what has been said in these first two chapters, the flaws in strategy may be glaring and obvious, but the questions which have to be addressed are:

● What did each of the three principals do (or not do) which contributed to the failure?
● What could they have done to prevent it?

The ultimate responsibility lay with the senior management of the company, but the blame was laid squarely at the feet of the maintenance manager. He was moved to another job.

3
Validating the work scope

Introduction

A turnaround is a task-oriented event and the list of tasks – the *work scope* – is the foundation upon which all other aspects of the event rest, especially safety, quality, duration, cost, resource profile, material and equipment requirements.

Initially there is no work scope, only lists of work requests generated by the production, maintenance, engineering, projects and safety departments and the like. At this stage the work may be either well or badly defined and it may or may not be necessary. The job of the plant and turnaround teams is to take these basic work lists and, using a technique known generally as *validation*, get the work requests properly defined and weed out the unnecessary work.

Unnecessary work

Why should there be unnecessary work? Surely the people generating the work lists know what needs to be done and would not complicate matters or generate extra costs by requesting work that was unnecessary?

To address this issue it is necessary to recall the principle, already referred to, of performing the absolute minimum amount of maintenance work commensurate with protecting the reliability of the plant. A turnaround is a maintenance event (in part) and so should conform to this principle. Work lists, however, are not generated by business managers – who have an eye on the bottom line, the profit margin – but by people who work on the plant (or are closely concerned with it) and who form certain personal judgements about the state of the equipment, the reasons for its various malfunctions, its chronic problems and what should be done to rectify these faults.

These judgements may be based on objective evidence and measurement but they may also be subjective, based on no more than a gut feeling. In either case, the conclusions drawn about causes of problems – and especially about what has to be done to put things right – may be faulty and should be tested against some objective criteria to ensure their validity and confirm the necessity for carrying out the work requested.

Another category of unnecessary turnaround work is that which could (and should) be done either while the plant is on line or during some other outage. A popular perception among plant personnel is that

The plant is shut down anyway, so why not get as much work done as possible?

Which seems at first glance to be a reasonable approach and is, in fact, one which is adopted in some companies. The main reason given by both maintenance and production managers is that they just do not have enough staff to carry out all the necessary tasks during normal operation of the plant. If this is so then it is a shortcoming in company staffing policy that is being masked by the turnaround. If work is being dumped into the turnaround then it is not being performed when it should be. This makes a nonsense of the maintenance programme. It should be recognized by the policy team and, if it is acceptable to them, a separate budget should be created and visibly labelled (for all to see) as 'Cost of work which should be done throughout the year but cannot be, due to a shortage of staff (or whatever other reason)'.

There are several ways in which unnecessary work impacts upon the turnaround and it is important to understand them if such work is to be challenged. They fall under the following headings:

(i) Expense of turnaround work
Because a turnaround requires a large control team, work done during the event is more expensive than that which is carried out at any other time of the year.

(ii) Unofficially transferred work
If work is transferred unofficially from the routine maintenance list to the turnaround work list it decreases the expenditure on routine maintenance (which is good for plant personnel) and increases the cost of the turnaround (which is not so good for the turnaround manager). Plant personnel who make such unofficial transfers of work often forget to transfer the funds set aside for the work in the everyday maintenance programme to the turnaround budget – and even if they did it would not cover the cost, because of reason (i). They may even forget to inform the turnaround team that the work could be performed at another time.

(iii) Overcomplicated work requests
Even if the work requested is necessary, the method requested may not be the most cost effective. For example, the request 'Remove 36" valve from top of column and overhaul' – no mean undertaking – could be reduced, after analysis by the turnaround team, to 'Re-pack valve in-situ', a far easier and much less costly alternative.

(iv) Duplication
Because work can be requested by a large number of plant and other personnel, two other categories of unnecessary work can arise, namely work from duplicated requests and work which is redundant. Duplication

arises when different people request the same job. On many plants there is a three-shift rotation system and on some this may involve up to seven shifts of personnel. Communication between the shifts and between shift and day workers may not always be effective.

Two requests for the same work can initially be mistaken for two different jobs because they have been written differently. Compare:

'Investigate fault on distillation column isolation valve and repair as necessary'

with the alternative request:

'Overhaul IV 3024'

If the work is not validated at the start, a lot of time and money may be wasted planning and resourcing unnecessary work.

(v) Redundant work

This is work that is requested but is made unnecessary by another work request which is not cross-referenced (see the case study at the end of this chapter).

(vi) Nice-to-do work

This is work for which there is no business justification, but someone, somewhere, is determined to get it done for their own comfort. As one plant supervisor said, 'OK, there was no indication that the cooling system was silting up but I thought – because the plant is shut down, anyway – it would be nice to take the end cover off the heat exchanger and have a quick look inside'.

The job was done and no problems were found. But a problem did arise when it came to boxing-up the exchanger. The fastening studs had been accidentally removed and scrapped and a new set had to be ordered. The supervisor got his 'nice to do' job done – at a cost of $3000!

(vii) Desperation work

This is the 'We have to do something!' category. Consider this: a large electric motor had run continuously for ten years without any trouble and there was no indication of any current problems. However, the electrical engineer was anxious that the motor would fail before the next turnaround (scheduled for two years ahead, exactly because it had run trouble free for so long!).

The motor was shut down, stripped and certain parts replaced with new – even though they showed no indication of excessive wear or damage. The unfortunate ending to this tale is that the motor burned out when it was re-started because the wrong type of bearing had been used to replace the original one. The duration of the turnaround was extended by two days and over half a million dollars' worth of production lost.

(viii) Blackmail work

This category normally occurs late on in the event, usually very close to start up when people are operating under severe time pressure and are on their last reserves of energy. It is presented by a member of the plant team who uses his local knowledge and the fear of duration over-run to force a task onto the work list.

A typical statement is 'If you don't do this, then...' followed by any number of dire conclusions, e.g. 'The plant won't start up', 'Production will be affected', 'Product quality will suffer'. All of these may or may not be true but the question has to be asked – if such dire consequences are attached to not doing the work – why was it omitted from the work list in the first place?

Unnecessary work wastes money and time, clogs up the schedule, diverts scarce resources from critical work and may introduce faults into an otherwise reliable plant. The worst case scenario is that the unnecessary job becomes the critical-path job (as in the case of the electric motor).

It is therefore vital that the raw work lists are subjected to validation, which should proceed via the following basic steps:

CHECK	– that the request has been approved by a manager or his nominee;
	– that the work is not duplicated or redundant;
	– that the data on the request are accurate and describe exactly what is being requested.
CHALLENGE	– the need for performing the task at all;
	– the need for performing the task during the turnaround;
	– the need for performing the task as requested (can it be done more economically?).
ANALYSE	– the safety/quality/material/equipment/resource requirements.
VALIDATE	– agree on the final wording of the work request;
	– record the work request on the approved work list.

The validation process is employed to ensure, as far as possible, that the approved work scope contains only what is necessary to restore, maintain or enhance the reliability of the plant and which cannot be done at any other time (see Figure 3.1, a validation routine used by an oil refinery).

An effective work scope is achieved by analysing every task, which is done at the following series of meetings.

The work list meeting

The effectiveness of the work list meeting will set the pace for all other activities on the turnaround. Speed, as well as accuracy, is important.

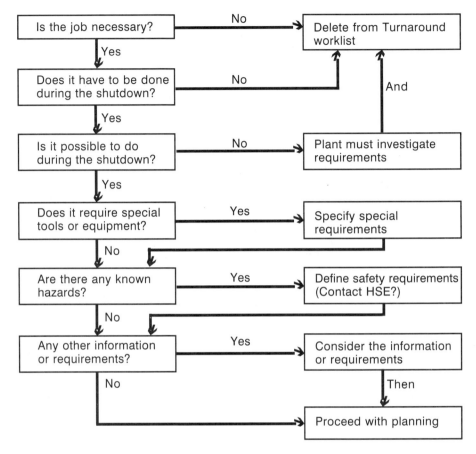

Figure 3.1 Sample work validation routine

Purpose

The meeting gathers together, and validates, the work requests generated by a large number of people in various departments, creating an approved work list for the turnaround.

Participants

Should include:

- Turnaround Manager (who is also the chairman);
- Preparation Engineer (if one is being employed);
- Planning Officer and relevant planners;
- Plant or Production manager;
- Maintenance Manager (or whoever is responsible for the maintenance function);

- Plant Supervisors;
- Electrical/Instrumentation Engineers and Supervisors;
- Nominees of either Plant or Turnaround Manager.

Work scope elements (see Figure 3.2)

A typical work scope would comprise the following categories of work:

- Statutory or company inspection requirements;
- Preventive maintenance;
- Corrective maintenance on known defects;
- Plant cleaning routines;
- Safety and quality initiatives;
- Work generated by the turnaround.

If there is a very large or complex work list the decision may be taken to split it into sections and conduct several meetings rather than a single general one, the work being divided by geographical area, functional unit or craft discipline. If this is done there must be at least one final meeting to draw all the sub-lists together into a single list for final validation.

Remit (see Figure 3.3)

The remit of the meeting is as follows:

- Meet weekly (or to another agreed schedule);
- Gather together every job request;
- Eliminate unnecessary work;
- Justify tasks which stay on the work list;
- Categorize all tasks;
- Clarify requirements for each task;
- Generate actions to drive the process;
- Assign the action to appropriate personnel;
- Formally minute all actions, decisions, progress and problems.

The major task review meeting

Major tasks must be properly planned and prepared, because at least one of them will define the duration of the turnaround and any of them, if not properly planned and prepared, may go wrong and become the critical path.

Purpose

The purpose of the major task review meeting is to ensure that large, complex or hazardous tasks are given the due consideration that their importance merits, so that a relevant specification for the task may be produced.

Check list of issues requiring consideration

Statutory/company requirements

- Inspect/repair pressure vessels
- Inspect/repair registered pipe work
- Inspect/repair relief rtreams/readers
- Overhaul/replace relief valves
- Inspect/replace bursting disks
- Inspect/repair of ultimate protection devices
- Inspect/overhaul non-return valves
- Inspect/replace bursting disks
- Inspect/repair suspect pipe work
- Pressure test to expose suspected leaks
- Inspect pipe work for suspected fouling

Repairs must be specified by the turnaround technical team and approved by the turnaround management team

Preventive maintenance

- Overhaul rotating machinery
- Overhaul electric motors
- Check equipment alignments and adjust
- Check instrument calibration and adjust
- Overhaul/replace control valves
- Inspect flameproof equipment and repair/replace
- Check vessel linings and internal equipment and if necessary repair
- Investigate suspected malfunctions
- Lubricate/change oil/top up oil reservoirs

Repairs must be specified by the turnaround technical team and approved by the turnaround management team

Corrective maintenance – known defects

- Repair tube or pipe leaks
- Repair flange and joint leaks
- Repair/replace steam traps/lutes/pots
- Repair furmanited joints
- Repair other temporary fixes
- Repair/replace - instruments - pumps
  - motors - gearboxes - valves - cable runs
- Repair/replace - handrails - access ladders
  - platforms - insulation
- Large asset repair
  - Compressors/turbines/large pumps
  - Vessel shells - internals - externals
  - Column trays and internals
  - Fin fan coolers - Cooling towers
  - Heat exchangers

Plant cleaning

- Cleaning requirements
  - Heat exchangers - vessels - pipelines
  - Equipment removed for overhaul
  - Cooling tower - Fin fan coolers
  - Drains - reservoirs - sumps - pits
- Cleaning techniques
  - Water washing - Jet washing
  - Chemical cleaning
  - Grit blasting/Slurry blasting
  - Manual cleaning
  - Mechanical cleaning
- Cleaning locations
  - In situ
  - On site water washing bays
  - On site decontamination bays
  - Off site cleaning facilities

Safety and quality initiatives

- Removing asbestos insulation
- Inspecting pipe supports
- Removing redundant plant
- Re-routing pipe work
- Special safety checks on equipment
- Complying with new safety regulations
- Upgrading equipment (modernizing)
- Planned gasket upgrading
- Upgrading foundations/roads etc.

Work generated by the turnaround

- Process washing and sweetening
- Isolating/de-isolating
- Safety checks prior to vessel entries
- Scaffolding - removing insulation - re-insulating
- Disposing of redundant plant and equipment
- Disposing of waste and hazardous substances
- Loop checks/trip and alarm checks
- Site cleaning and housekeeping
- Peripheral tasks
  - Interfacing with other production units
  - Working on line (stoppling etc.)
  - Battery isolations
  - Controlling shared services

Figure 3.2 Work scope elements

Check list of issues requiring consideration

Gather together every job request

- Ensure all work lists have been submitted
- Statutory inspection – Company inspection
- Major tasks – Plant modifications
- Project work – Machine overhauls
- Preventive tasks – Corrective tasks
- Equipment cleaning – Catalyst/packing
- Instrument tasks – Electrical tasks
- Safety initiatives – Quality initiatives
- Peripheral work – Plant investigations

- Freeze the work list at the specified date

Eliminate unnecessary work

- Challenge client's practices
- Challenge every job - does it need to be done - or done this way - or done so frequently?
- Ask - What are the consequences of not doing this job? If the answer is 'none' eliminate it
- Defer work to smaller outages
- Eliminate extremely hazardous work or change the requirement to make it safer
- Eliminate unauthorized or duplicate work
- Eliminate 'nice to do' work unless the plant manager authorizes and accepts the costs

Justify jobs which stay on the list

- Ensure the job is technically feasible
- Ensure job is economically feasible
- Check due dates for all statutory work
- Ensure the job can be done in the turnaround
- If the job is a modification, ensure the requisite senior management approval is obtained
- If the job means working on live plant, ensure senior management approval is obtained
- If a job is likely to take longer than the fixed duration of the turnaround, inform policy team

Categorize work

- Categorize as follows
  - Major tasks – to be worked up by major task review team and planned by preparations engineer
  - Small tasks – to be planned by planning team and be given a separate job method sheet
  - Bulk work – to be planned by planning team and be packaged on schedule lists
- Make out three separate work lists

Clarify requirements for each job

- Is there any information to indicate the job has abnormal technical or safety features?
- Are there any special techniques that have to be employed to do the job?
- Do the plant management require the job to be done in a particular way due to local needs?
- Is there any evidence to suggest the job has been difficult to do in the past or needed any special techniques or equipment?

Generate actions to drive the process

- Don't leave anything to chance or guesswork
- A great deal of information not on the work orders will be needed to specify the task
- Clarify the needs into specific actions in order to progress the job, such as:
  - Request information from a known source
  - Find out where information can be obtained
  - Reorganize information into required format
  - Investigate or measure up on-site
  - Formulate a plan or strategy

Assign actions to appropriate person

- Ensure the person who is assigned the action is capable of carrying it out
- Ensure the action is specific
- Agree an end date with the assignee
- If the assignee requires any special resources to carry out the action, provide them
- At each meeting, review outstanding actions and if they are not completed, find out why!
- Take all reasonable measures to ensure actions are completed on time

Other considerations

Figure 3.3 Work list meeting remit

Participants

Should include:

- Preparation Engineer (the chairman);
- Plant or Production Manager;
- Maintenance Manager;
- Safety representative;
- Other nominated plant personnel;
- Any co-opted for their specialist technical knowledge.

Remit (see Figure 3.4)

- Meet at agreed intervals until all issues are resolved;
- Obtain all information and documents necessary to plan the task;
- Describe the plant or equipment in detail;
- Define the task, step by step;
- Identify and resolve technical challenges;
- Carry out a task hazard study;
- Formulate a safety routine and a rescue plan;
- Define the following:
  - A preparation plan;
  - Inspection requirements;
  - Material, equipment, services and utility requirements;
  - Manpower requirements;
- Minute all actions, decisions, progress and problems.

The inspection review meeting

Because, in some countries, various inspections are required by law, it is vital that the tasks be specified in such a way as to achieve the inspectorate's objectives and that the specifications are auditable (because the inspectorate may wish to examine them).

Purpose

The inspection review meeting is held to ensure that all necessary statutory and internally generated inspection requirements are identified and defined – and met via the employment of the most relevant techniques.

Participants

Should include:

- Preparation Engineer;
- Planning Officer and relevant planners;
- Chief Inspector (optional);
- Inspectors assigned to the overhaul;
- Any other nominee.

Check list of contingencies to be considered

Information and documents
- Work order number
- Plant history of the equipment
- Line diagrams and P and IDs
- Inspection, test and IMI reports
- Operating information
- Process information
- Assembly and detail drawings
- Past turnaround reports
- Accident/incident reports
- Any other useful information etc.

Plant details
- Plant name and number
- Type of plant
- Fabrication materials
- Process and/or services fluids
- Temperature/pressure/flow
- Access or egress hazards
- Entry problems/requirements
- Residual substances
- Any known hazards
- Any other useful information

Task to be done
- State reason for task
- Produce a step by step plan
  - Main task stages
  - Initial logic network
  - Activities for each step
  - Duration of each step
  - Timescale for complete task
- Define if task is critical path
- Produce an overall plan

Technical challenges
- Erosion/corrosion/damage/distortion/misalignment
- Design or project input
- Special material (long delivery)
- Special welding or machining
- Special skills or knowledge
- Specialist sub-contractors
- Special analysis techniques

Task hazard study
- Identify potential hazards at each stage of the task
- Define the potential loss which is associated with hazards
- If possible, eliminate hazard
- If elimination is not possible, take steps to guard against hazards

The safety routine
- Define special requirements:
  - Inert entry procedure
  - Working in confined spaces
  - Working at height/multi-level
  - Hot work/work on live plant
  - Multi level working
  - Need for entry guardians
  - Communication needs
  - Standby emergency services
  - Safety equipment

The rescue plan
In case of an accident on a work site with difficult access/egress
- Ensure there is a rescue team
- Alert the rescue team before work starts
- Ensure there's an escape route
- Ensure an inert body can be removed from the job site
- Ensure standby emergency services

Preparation plan
- Shut down/start up programme with durations specified
- Cleaning - washing - cooling
- Isolation requirements
- Ventilation required
- Removal of insulation
- Safety monitor and test
- Access and platforms
- Permits to work

Inspection needs
- Previous inspection reports
- Company/external inspectors
- Inspection technique (special)
- Timescale for inspection
- Platform – moveable/modified
- Lighting/sheeting out
- Calibrated instruments
- Radiography/other hazards
- Statutory requirements

Materials
- Specification
- Traceability
- Availability
- Long delivery
- Consumables
- Delivery/protection/storage/lay down areas/issue
- Special needs (welding etc.)

Equipment
- Cranes/special handling gear
- Generators/compressors
- Wagons/low loaders/bogeys
- Disposal gear (catalyst etc.)
- Scaffolding, ladders etc.
- Special tools (torque gear etc.)
- Pressure testing equipment
- Delivery/lay down areas
- Servicing (if required)

Services/utilities
- Electricity/gas/water/fuel
- Water wash/grit blasting/chemical cleaning
- Lagging/insulation/painting
- Cleaning/substance disposal

Manpower
- Number of people required
- Disciplines/trade split
- Training/briefing needs
- Experience needs
- Special skills or knowledge
- Company/external manpower
- Working hours/shift patterns
- Relief workers/stand-ins
- Special payments
- Industrial relations issues

Other requirements

Other requirements

Figure 3.4 Major task review meetings

Remit (see Figure 3.5)

- Meet regularly until all issues are resolved;
- Create a categorized inspection list;
- Explore the possibilities of deferring inspection (if desirable);
- List all inspectors' requirements for each job;
- Define inspection techniques to be used;
- Define types of inspectors required;
- Minute all actions, decisions, progress and problems.

Project work review meeting

Although project work is performed on the same plant at the same time as the turnaround, it may be controlled by several different departments. It is therefore vital that it is properly integrated into an overall schedule, identifying interactions with turnaround work and avoiding conflicts.

Purpose

The meeting should ensure that any project work, which is normally planned by the project department, is properly integrated into the turnaround schedule so that all conflicts are resolved and requirements met.

Participants

Should include:

- Turnaround Manager;
- Planning Officer;
- Plant Manager;
- Project Manager (optional);
- Project Engineers;
- Any other nominees.

Remit (see Figure 3.6)

- Meet regularly until project issues are resolved;
- Specify major and minor projects;
- Clarify key dates;
- Review documentation;
- Review material and equipment procurement;
- Define interactions with turnaround work;
- Minute all actions, decisions, progress and problems.

Check list of issues requiring consideration

Categorized inspection list

Should be categorized as follows:

- Pressure vessels
- Relief streams
- Registered pipe work
- Nominated pipe work #
- Registered non-return valves
- Ultimate protection devices
- Any other nominated inspection #

 # Plant manager nominates

Deferred inspections

In certain circumstances, a case can be made for deferring inspection, if:

- Past inspection indicates no appreciable deterioration in condition
- Evidence from other plants in the same industry sector indicates no appreciable deterioration
- On-line monitoring indicates that there is no appreciable deterioration
- New on-line techniques make it no longer necessary to open up vessels etc.
- Plant performance or physical data shows that the element has not deteriorated

This option should be considered

Inspector's requirements

In order to carry out the job the inspector will need some combination of the following:

- Vessel atmospheric clearance certificate
- Permit to work
- Guardians/communications
- Assistants
- Protective clothing
- Safety gear
- Breathing apparatus
- Working platforms/access ladders
- Lighting/electric power
- Ventilation/fans/air movers
- Briefing on any special circumstances

Inspection techniques

Various, including, but not limited to:

- Visual inspection
- Dimensional measurement
- Dye penetrant
- Magnetic particle
- Ultrasound
- X-ray or Gamma radiography
- Bubble testing
- Thermal Imaging
- Photogrammetry
- Other specialized methods

Types of inspectors

Due to the various types of work done, a number of different types of inspectors may be needed, including but not limited to:

- Pressure vessel inspectors
- Welding inspectors
- Radiographers
- Registered pipe work inspectors
- Machine inspectors
- Specialist technique inspectors

Other considerations

Figure 3.5 Inspection review meeting

Check list of issues requiring consideration

Types of major projects

Various types of major project include, but are not limited to:

- Plant extension
- Major asset renewal
- De-bottlenecking
- Major plant modifications
- Major safety modifications
- Process Improvement initiatives
- Large construction activities
- Large civil activities

Value typically more then 150,000 US dollars

Types of minor project

Various types of minor project include, but are not limited to:

- Minor asset renewal
- Fixed cost reduction
- Safety initiatives
- Quality initiatives
- Packaged work
- Small design packages

Value typically less than 150,000 US dollars

Key dates for major projects

To ensure project is ready in time the following key dates should be conformed to:

Key project activity	Time before turnaround
– Initial project definition	18–24 months
– Initial project meeting	12–15 months
– Meeting with turnaround team	6–9 months
– Contract awarded	6 months
– Organize pre-shutdown work	6 months
– Project closure date	2 months
– Project plan issued	2 months
– Delivery of all material and items	1 month
– Briefing of project work force	1 week

Project documentation

To ensure everyone concerned understands the project requirement, review the following:

- Feasibility/operability study
- Modification approval document
- Design and detail drawings
- Line diagrams
- Break-in drawings
- Materials lists and delivery schedules
- Equipment lists and delivery schedules
- Contractor documents
- Pre-shutdown work list
- Isolation register
- Project schedule for integration into turnaround plan

Review material and equipment procurement

It is essential that the correct material/equipment (M and E) is on site when it is needed, so ask:

- Has all long delivery M and E been ordered?
- Who is responsible for expediting M and E?
- Who will take delivery of M and E?
- What are the M and E storage requirements?
- Who will inspect M and E to ensure conformance?
- Will any M and E arrive after plant shutdown?
- Does any M and E require special handling?
- Is any M and E hazardous?
- Will contractors bring their own M and E?

Interactions with turnaround work

- Who will manage the project?
- Will the project be the critical path?
- Can the project be packaged with any other turnaround work?
- Is there any conflict between the Project and any other turnaround work?
- Will the project require the involvement of anyone already dedicated to the turnaround?
- Will contractors bring their own management, inspection and supervision?
- Will any specialist vendor require assistance from any turnaround personnel?
- Who will be responsible for removal of the redundant M and E generated by the turnaround?

Figure 3.6 Project work review meeting

The validated work scope

All other aspects of turnaround planning, i.e. safety, quality, costs, materials, equipment and resource requirements are derived from the work scope. It is not possible to plan against an open-ended work list. Therefore, once the scope has been validated the work list should be frozen on a date agreed by the policy team. Thereafter, any request for work must be authorized by the highest executive authority available and, if it can be justified and is approved, should be placed on a 'late work' list. The money, time and resources required to carry it out are extra to budget and are to be identified as such. Typically, there will be a steady flow of late work requests throughout both the preparation and execution stages of the turnaround (those arising during the latter stage are for 'emergent work', which is discussed in a later chapter). A validated work scope simplifies planning and preparation; late work complicates them.

Case study

A turnaround work list for a chemical plant contained a request for the replacement of two double-block-and-bleed valves on an eight-inch line, the existing valves being unserviceable. The work was validated and planned, and new valves bought – at some considerable cost. On Day 4 of the turnaround the two unserviceable valves were removed and replaced by two new valves. The job took two days to complete. On Day 6 (just as the previous job was nearing completion) the eight-inch line was cut out and replaced by a ten-inch line as part of an upgrading project – complete with two ten-inch double-block-and-bleed valves, brand new.

An investigation revealed that, although the project programme had been integrated into the overall event schedule from a timing, duration and resource point of view, the technical content of the job had not been discussed at any turnaround meeting. The operations supervisor who had been involved in the definition of the up-grading project, and therefore knew that the eight-inch line would be removed, had not passed this information on to the senior operator who had requested that the eight-inch valves be replaced. The question is – what safeguards can be put in place to prevent occurrences like this?

The ultimate responsibility lay with the turnaround manager. If he had ensured that all projects were technically defined at a meeting which was attended by some of the same people who attended the work list meetings, the duplicated work would have been identified.

4
Pre-shutdown work

Introduction (see Figure 4.1)

Although the main business of the preparation phase is to plan the work which will be performed *during* the event, part of the preparation effort must be dedicated to ensuring that all work which has to be completed *beforehand* (if the event is to be performed effectively) is also planned and executed. It must be identified, progressed and completed as indicated in the upper section of the preparation network shown in Chapter 1 (see Figure 1.2).

Long delivery items

A significant part of the planning and preparation of a turnaround is the timely procurement of hundreds, or even thousands, of items – materials, spares, proprietary plant equipment and so on. Some of these – e.g. very large components of critical machines or parts made from exotic materials – will be on long delivery (as mentioned before, a delivery time for a compressor rotor might well be as much as sixteen months). In order to identify these, it is necessary to analyse the work list as early as possible to ensure sufficient time is allowed for the various procurements. Remember, that although the delivery time quoted is from receipt of order, the order will firstly have to be transmitted through the company's procurement system – which has its own time scale that can sometimes run into weeks, or even months. Also, after

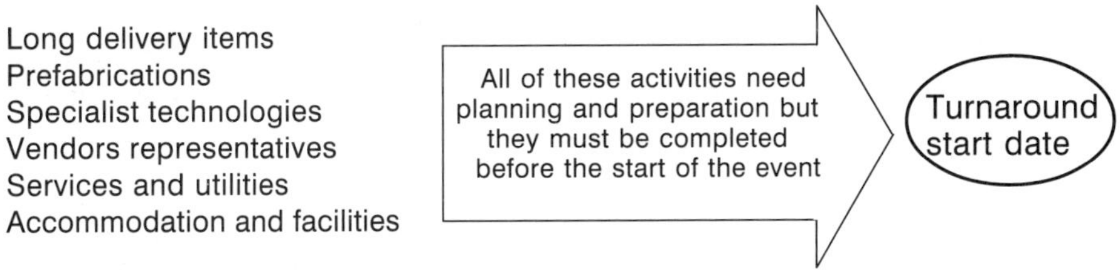

Long delivery items
Prefabrications
Specialist technologies
Vendors representatives
Services and utilities
Accommodation and facilities

All of these activities need planning and preparation but they must be completed before the start of the event

Turnaround start date

Figure 4.1 Pre-shutdown work

delivery on site there will normally be a requirement for goods inward inspection – and if the item fails this it will have to be replaced!

What to look for

When analysing the work list, categories which should flag up the need to check on delivery time at an early date include, but are not limited to:

- Rotating equipment or spares;
- Large or specialized valves;
- Non-standard equipment;
- Large quantities of non ex-stock items (eg reformer tubes or column trays);
- Exotic materials;
- Catalyst and packing;
- Items procured from foreign countries;
- Items needing special permits or certificates;
- Items requiring further work after delivery (assembling, painting etc.).

What to do

When long delivery items are identified the following procedure should be followed:

1. Make out a list of the items.
2. Find out if they have already been ordered.
3. If not, identify who will procure them and then instruct him to:
   (i) check whether promised delivery is before the turnaround (if it isn't to inform you of the date of delivery during the turnaround);
   (ii) expedite and regularly report on progress and problems;
   (iii) (if practical) witness any vendor's tests and inspect item before it is delivered.
4. Arrange for storage or lay-down area (which may be on site or remote) and ensure the item will then be protected and maintained.
5. Arrange for any goods inward inspection and quick return of any damaged item to the supplying vendor.
6. If item will not be delivered until after the start of the turnaround, calculate when it will be available for installation.
7. Report any major problems to the policy team (e.g. delivery times which will mean that installation of the items concerned will extend the duration of the turnaround).

It will be up to the policy team to decide whether the work should be eliminated from the turnaround list or modified, or whether the criticality of the task would justify the extension of the duration. However, even after that decision has been taken, one should either work to bring forward the delivery date, modify the schedule to accommodate the delivery time or specify a different task which will delay the necessity for installing the item (if the replacement is going to be late, can the damaged item be refurbished to last a few more months until the replacement can be installed?).

Prefabricated work

This is usually pipework or structural steel and can cause problems similar to those posed by long delivery items. In this case the materials or items which make up the fabrication may be on short delivery or even ex-stock. What takes up the time is the fabrication itself. The actual prefabrication work may be done on-site by local or contract manpower or it may be done off-site by a company contracted to do the work.

Prefabrication may have no time pressure on it whatsoever and that in itself can be a problem. Due to the belief that there is plenty of time, it is constantly put off until it *does* become a time pressure problem. Therefore, the turnaround manager must ensure that all prefabrication is carried out in a timely manner.

What to look for

- Large amounts of small fabrications (steam traps etc.);
- Large single fabrications;
- Fabrications of exotic materials;
- Fabrications with complex configurations;
- Fabrication which will have to withstand high pressure or load.

A list should then be drawn up of all special requirements, such as:

- Non destructive testing;
- Pressure testing;
- Linings, coatings or cladding.

What to do

1. Obtain a copy of the prefabrication documentation, i.e. isometric or other drawings, welding procedures, pressure testing procedures etc.
2. Check whether anything on the prefabrication bill of materials is on long delivery (if so, use the procedure described in Chapter 5).
3. Double check the fit. Get the planner to go on site and measure up the job to see if the configuration of the prefabrication on the drawing will actually fit into the available space. Designers often work from site drawings which may not show what is actually there.
4. Check all testing and inspection requirements and ensure the equipment needed will be on hand and properly calibrated (with accompanying

certificates). The equipment itself may have to be procured and it may be on long delivery!

5. Check on the materials and items for fabrication, to identify those which require material or conformance certificates. Check (before delivery) that the certificates have been requested from the vendor and (after delivery) that they have been provided and are correct.

As with long delivery, major problems must be reported to the policy team so that they can be resolved.

Specialist technologies

As the technologies of maintenance become ever more sophisticated they increasingly involve expensive equipment needing more specialization in their application. A large number of specialist companies have emerged to perform the necessary work. Although the rates they charge are high, the various applications of their technologies can save much time and money and are consequently worth considering. The following list is representative of these, but is by no means exhaustive:

- Photogrammetry;
- Thermal imaging;
- Remote cameras, videos, introscopy;
- Inert entries;
- Bolt tensioning and torquing;
- Laser alignment and measurement;
- Sound wave signature measurement;
- On site heat treatment;
- Arc air/plasma/water jet cutting;
- Metal spraying;
- Application of ceramic metals;
- Reformer servicing;
- Vibro-testing;
- Mobile gasket workshops;
- Mobile water and jet washing facilities.

These technologies are normally provided by small companies or small divisions of large companies. There may be no relevant specialists in the particular country where the turnaround is taking place, so they would have to be brought in from elsewhere with all the attendant difficulties of doing so. Because there are not many companies in any one field they are normally booked up many months in advance and do not have the flexibility to suddenly change either their manning levels or procedures to meet a particular demand. It is therefore very important that they are contacted as early as possible and given accurate specification of the work required. If possible, they should be asked to come to the site and demonstrate their technology so that it may be better judged whether they are appropriate to the particular need. Alternatively, another site where the technology is being applied could be visited to inspect it in action.

Once the decision to use the technology has been made, the routine (described in Chapter 5) for handling contractor packages should be followed.

Vendors' representatives

There are certain plant items – such as boiler packages or digital control systems – which are normally serviced by vendors' representatives. This is frequently linked to warranty and in some cases there is no alternative but to use them – and they will normally have a heavy workload. In order to ensure that they will be on hand when required (i.e. during the turnaround), and can provide a good service, the following steps should be taken as early as possible during the preparation phase:

(1) Check their availability during the turnaround.
(2) If they are not available, request that someone else be recommended.
(3) Ask them if they are familiar with the plant concerned.
(4) Inform them of any current plant faults.
(5) Check that they have the necessary materials.
(6) Find out what support they will require.
(7) Acquire a work scope and duration from them.
(8) If it is feasible, ask them to visit the site before hand to check the equipment before the turnaround starts.

Services, utilities, accommodation and facilities

Although the management of these requirements will be discussed in detail in Chapter 8, it is timely to state here that all services (water washing, scaffolding, lagging etc.), utilities (water, electricity and gas), accommodation (offices, rest rooms, stores etc.) and facilities (canteens, washrooms, toilets etc.) must be in position on site and serviceable before the turnaround starts – so the planning, resourcing and execution of this undoubtedly qualifies as pre-shutdown work.

Case study: Red Zone

As part of the pre-shutdown preparation work for a very large turnaround on a chemical plant, the management addressed the problem of long delivery items by creating a very simple but effective material 'Red Zone' routine to handle the many thousands of items required for the turnaround – many of which were procured from abroad.

Firstly, the 'Red Zone Philosophy' was established. This stated that any item which was promised for delivery within four weeks of the start of the event (some items were not due to be delivered until after the start date) would be declared 'Red Zone'. These had to be handled by an appointed expediter, whose instructions were as follows:

(1) Contact all suppliers or manufacturers of items in the Red Zone and, if possible, negotiate a new delivery time which will remove the item from the zone.

(2) Agree, with each manufacturer, a progress and delivery programme with key dates which can be monitored.

(3) Ensure the manufacturer understands the paperwork which will be required to accompany the item (test certificates etc. – these were listed).

(4) Regularly monitor the progress of each Red Zone item (in some cases this may require a visit to the vendor's premises) to check that key dates are being met and that the promised delivery date will therefore be achieved.

(5) Ensure that, as far as is practically possible, all required inspections or tests are carried out at the manufacturer's premises and, if necessary, appoint an independent inspectorate to represent the company (especially in foreign countries).

(6) Alert the relevant engineer at the earliest date of any issue which is likely to cause a delay in the delivery of the item.

(7) Check, just prior to delivery, that all paperwork has been generated and will accompany the item (or, if sent under separate post, will arrive before the item).

(8) Negotiate (or appoint an agent to do so on behalf of the company) with customs officials in both the exporting and the importing countries to ensure that all exporting and importing regulations are complied with, thus ensuring that this issue will not delay delivery.

(9) Check the entire delivery route to identify any potential hold ups.

(10) At the earliest date, identify and calculate costs of alternative delivery routes for any item facing a potential delay – this to include an emergency system for items delayed at the last minute.

(11) Sign off the item as delivered only when all goods inward checks have been completed and the resulting documentation is in hand.

(12) Record all issues which caused – or which might cause – delays and make recommendations for expediting future programmes.

None of this is rocket science! It is merely meticulous attention to detail – attention which leads to the overall success of a programme of procurement of turnaround items.

At the end of the event the expediter's final report drew attention to several cases where, if the Red Zone had not been in operation, items would have arrived late, and two instances – both of which concerned large critical items – where the item would not have arrived at all. Oddly enough, the report was never published, so although the lessons were learned on an individual basis, they were not added to the sum of corporate knowledge.

5
Contractor packages

Introduction

It is not within the scope of this book to discuss the work of a company's procurement or contract departments. Most organizations have a well-developed strategy for organizing the contractor requirements for any project. However, the unique nature of a turnaround project demands additional action by the turnaround team. This requires, in turn, that the team has a basic level of commercial awareness and a working knowledge of the main issues surrounding the use of contractors.

In the context of turnarounds 'contractors' are companies that specialize in performing some or all of the tasks and activities involved in an overhaul. They may be very specialized (e.g. an electronics company servicing a digital control system), multi-skilled (able to carry out mechanical, fabrication and civil work), or may specialize in the total planning and execution of events.

Whichever type of contractor is engaged, they are all commercially aware companies who are in business to make a profit and, because of greater familiarity with the business of turnaround contracting, will be much more skilled at negotiation than the client company. The majority of them are reputable companies who make a valuable contribution to the business of turnarounds and will perform work effectively so long as all the necessary preparations have been made by the client.

In many of those cases where a client company feels it has been 'ripped off' by a contractor the truth of the matter is that the extra money it had to pay was the penalty for its own bad work specification, disorganization during the event and lack of the basic commercial awareness that, contractually, some things are possible and some things are not. The contractor will honour the contract but he will also demand payment for all the work he is asked to do – planned and unplanned.

Using contractors – the upside

There are several reasons why contractors are used, and the following list (which is by no means inclusive) gives some of the main ones.

Resource availability
The planning and execution of a turnaround typically employ more people, and demand the availability of more skills, than are required just to operate a plant – a situation which is remedied by engaging contractors.

Experience
Even where manpower is available on the plant, the fact that it is only shut down once every two, three or even five years means that personnel do not get the opportunities to practice the skills required for executing turnarounds and, as the Americans say, 'if you don't use it you lose it!' Contractors, on the other hand, are doing the same work week in and week out and should be more skilled at performing the required tasks, but *caveat emptor*! (Let the buyer beware!) – for reasons that will shortly be made clear (see 'Using contractors – the downside').

Professionalism
As mentioned earlier, many contractors concentrate on doing one or two things and, in consequence develop a very high quality service.

Specialism
Some work is so specialized, complex or hazardous that there is no alternative but to use a specialist contractor or a vendor.

Productivity and cost
Audits indicate that good contractors provide higher productivity at lower cost to the client company than it could achieve with its own labour. Also, it is easier to replace sub-standard contract workers.

Using contractors – the downside
There are also disadvantages to using contractors.

Unavailability when required
Contractors will also be required by other companies for similar work. A total or preponderant dependence on contractors may therefore give rise to problems if the procurement system is not of the highest calibre.

Contractors not always as effective as their proposals suggest
One of the reasons for this arises out of the above unavailability problem. If the contractor is successful his resources may become stretched. If he cannot cover the contract with his normal personnel he has a problem. Because he doesn't want to lose the work he may provide some of his key staff and then supplement them with people 'off the street' who are of unknown and variable quality – hence the earlier *caveat emptor*.

The 'fitter' you are supplied with may well be a carpet fitter and the 'foreman' may have been driving a bus the previous week.

One factor should be borne in mind. The contractor's proposal will invariably have been based on the performance to be expected of a full strength team. If he fields a second string the client doesn't get the quality paid for.

Safety performance

If the contractor's safety systems are sub-standard, or if the men provided by the contractor are not well-practised at working on hazardous sites, the accident rate can be very high. The safety performance of contractors is, in some cases, not comparable with that of the client (although, it must be said, sometimes the reverse is true).

Difficult client–contractor relationship

A difficult relationship may terminate, in extreme cases, in a conflict which can only be resolved via litigation, which usually centres on claims for payments for delays, changes of intent and emergent work, but can also be due to the fact that the client or the contractor (or both) underestimated the volume or difficulty of the work.

Conflict between contractors

If many contractors are used (on major turnarounds they could number more than twenty) there will be problems at the interfaces between them – the most common are:

- One contractor delaying another.
- Two contractors trying to work in the same area at the same time.
- One contractor undoing another one's work.

For example, the welding contractor completes a number of welds on a pipe by eight o'clock in the evening. Radiography has been booked overnight. The following morning the lagging contractor (in conformance with the schedule) insulates the pipe. Later that day he finds the lagging has been stripped off. Due to a hold-up on another job the NDT team hadn't radiographed the welds overnight and the insulation had to be removed to allow the late radiography of the welds. Whose fault is it? Who should pay for the re-work? These are problems of co-ordination and are ultimately the responsibility of the turnaround manager.

Agency labour (body shops)

One difficulty which can constitute a *force majeure* may arise if the contractor hires mainly agency labour. Because turnarounds are of short duration and therefore often regarded as 'casual work' agency labour will always be on the lookout for new work. If there is the offer of longer or more lucrative employment at another location, such manpower may quit the turnaround without warning and head for greener pastures, leaving the client with insufficient labour to complete the turnaround within the planned duration – and in the short term there is nothing that can be done about it. The company may be especially vulnerable when the labour which disappears is a scarce resource such as coded welders or instrument technicians.

With the above in mind, it would perhaps be stating the obvious to say that much care and attention is required in the creation of contractor packages and the selection of contractors. Bitter experience teaches most turnaround managers that this is a difficult and demanding task and the message needs to be constantly re-emphasized to the turnaround team. Using contractors will not solve all the company's problems. Rather, it will exchange one set of problems for another.

Contractor work packages

The main factors influencing the selection of contractors are:

- the work scope and how it is packaged;
- the design of the turnaround organization;
- the type of contract to be awarded;
- the availability of contractors.

Because they are inter-connected they also influence each other.

There are a number of specific options for packaging work, i.e. by work type, functional unit, geographical area or contractor availability. In reality, the most likely situation will be that a combination of the above options is integrated into an overall contractor plan.

By work type

A given work package is created by grouping similar tasks together and awarding the whole to one contractor to make it more manageable and more economic to perform. The most common packages are as follows (a representative but not exhaustive list):

- Valve overhaul and replacement;
- Pump overhaul and replacement;
- Pressure vessel inspection;
- Re-traying columns;
- Catalyst and packing;
- Machines overhaul;
- Instrumentation;
- Electrical work;
- Welding (especially coded welding);
- Water washing and cleaning;

- Scaffolding (or staging);
- Lagging (or insulation);
- Painting (or protective coating).

The well-defined package forms the work scope for the contractor. Because each package is made up of similar repetitive tasks it is simpler to price, and its execution easier to control. On the other hand, the resulting engagement of a number of contractors can give rise to contractor–contractor conflict.

By geographical area

The plant is divided into discrete areas and all of the work (apart from highly specialized or hazardous tasks) in any one of them is given to one main contractor. While this arrangement can be complicated, due to the mixture of different tasks that the one contractor must perform, the advantage is that, having only one contractor in the particular area, there is no need to manage conflicts between contractors.

By functional unit

One contractor is awarded responsibility for a whole steam system, for example, undertaking many different tasks across a number of geographical areas. The obvious advantage is a greater guarantee of resulting system integrity because it is under a unified control. Conversely, if several contractors work on different parts of one functional system the turnaround manager must provide a co-ordinator to ensure all interfaces between contractors are properly managed and no work is left undone (because one contractor mistakenly believes that one of his own tasks is part of another's work package).

Contractor availability

Contractor availability cannot be guaranteed. It is unwise to take it for granted and to expect the contractor to be available – on short notice – just when you need him. Availability depends upon a contractor's workload and this in turn depends upon a number of factors such as:

- the time of year (turnarounds occur cyclically – fewer are done in winter);
- the concentration of turnaround-dependent industry in the particular area;
- the type of work involved (the more specialized, the fewer people do it);
- the amount of work involved (major events may require thousands of men).

A turnaround manager has to protect the interests of his client. Giving due regard to the above factors, he must adopt a strategy to do so, one which would include, *inter alia*, the following elements:

(1)   Finding out when other companies in the area will be undertaking turnarounds;
(2)   Identifying windows of opportunity to execute the overhaul;
(3)   Compiling a list of contractors and auditing their suitability;
(4)   Determining the availability of suitable contractors at the time of the turnaround;
(5)   Inviting competitive tenders. (Sending out invitations to bid (ITBs) as early as possible);
(6)   Booking specialist contractors as early as possible.

and, most importantly, securing the future by building long-term relationships with high quality contractors.

Types of contract (see Figure 5.1)

There are a number of different types of contract which can be awarded. The award will depend upon the circumstances surrounding particular packages of work. The types of contract are defined separately below, but in reality a complex project such as a turnaround will consist of a combination of any (or all) of the options shown. They are:

- Single contractor managed contract;
- Management fee and re-imbursed labour cost;
- Fixed-price packages;
- Call-off contracts (on scheduled rates);
- Day-work rates.

Each of these types of contract has its own features, benefits and drawbacks (see Figure 5.1).

When awarding contracts there must be a clear understanding of all the circumstances surrounding the turnaround from the perspective of:

- Contractor resource availability;
- Time-frames and windows of opportunity;
- Available finance;
- Different types of work required;
- Complexity of work;
- Need for specialized handling;
- Ability to manage the different types of contract.

Once the types of contract have been chosen, the next step is to select the contractor.

Contractor selection (see Figure 5.2)

In order to maximize the likelihood of the contractor performing effectively, the turnaround manager should consider the following actions:

Check list of issues to be considered

1. Single contractor managed contract

- *Features*
  - Requires a well defined, validated work scope
  - One contractor is awarded the total work scope
  - Contractor bids a fixed price for the contract
  - Contractor plans and co-ordinates all work
  - Contractor may execute work or sub-contract specific packages to other companies
  - Client facilitates contractor effort and monitors progress and performance
- *Benefits*
  - Contractor manages the total event
  - Contract cost and duration are fixed
  - Contractor must conform to agreement
  - Contract variations can be handled easily
- *Drawbacks*
  - Expensive preparation period required
  - Client needs excellent commercial skills
  - Client pays for all delays and downtime
  - Contractor's may underbid to win the contract and then try to generate extra work
  - If duration is tight, contractor may cut corners and compromise safety and quality
  - Client and contractor may become adversaries

2. Management fee and reimbursement

- *Features*
  - Contract agreed before work scope finalized
  - Contract executed to a fixed duration
  - Contractor manages and plans the event and provides labour to carry out all work
  - Contractor may subcontract work packages
  - Client pays a fixed fee for management and reimburses contractor for labour hours expended
  - Client monitors and validates contractor man-hours
- *Benefits*
  - Does not require early validation of work
  - Rates offered by different contractors can be compared and the best deal negotiated
  - Client can negotiate productivity agreements
  - Contractor bears the cost of over-run
- *Drawbacks*
  - Encourages slack planning by the client
  - Total cost will not be known until after the event
  - Client must closely monitor contractor man-hour claims to avoid 'padding out'
  - Work content cannot be controlled as well as on a fixed price contract
  - Work may overwhelm resources

3. Fixed price packaged contracts

- *Features*
  - Similar work grouped into a package
  - Contractor bids a fixed price for the package
  - Contractor manages the package
  - Client manages the contractors
- *Benefits*
  - Costs and duration are fixed
  - Onus on contractor to work effectively
  - Usually very few variations
- *Drawbacks*
  - Client pays for all delays and downtime
  - A number of contractors whose needs conflict
  - Client needs excellent co-ordinating skills

4. Call off contracts (scheduled rates)

- *Features*
  - Used for work that is difficult to specify exactly but can be categorized (scaffold etc.)
  - Works cope does not need to be defined
  - Client manages the contractors
- *Benefits*
  - Once agreed, all tasks are done at unit cost
  - Variations handled at unit cost
  - Client can negotiate productivity deals
- *Drawbacks*
  - Requires quantity surveying by client
  - Call off system is relatively easy to abuse
  - Can generate daywork with resulting costs

5. Daywork rates contract

- *Features*
  - Manpower hired to perform unspecified work
  - Contractor or client may supervise
  - Usually for general work only
- *Benefits*
  - No planning required – short lead time
  - Manpower more readily available
  - Easier to vary numbers at short notice
- *Drawbacks*
  - Very variable skill level – usually low
  - Client has to manage labour – high costs
  - Tend to be hidden costs and expenses
  - Difficult to achieve good productivity

Other considerations

Figure 5.1 Types of contract

Before the event

- If possible, use a known, high quality, contractor (If contractors are unknown quantities, invite at least three to bid for each package).
- Draw up a short list of companies and if possible, audit and compare them.
- Send ITBs out early to give the contractors sufficient time to assemble their bids.
- In all ITBs, clearly specify:
  – an accurate work scope (what is required, to what standard);
  – any special requirements (e.g. the need for qualified coded welders);
  – any special circumstances (e.g. working next to live, hot plant);
  – any built-in non-productive time (e.g. awaiting permits to work).
- When the contractor's bid are received , evaluate them and negotiate improvements to unsatisfactory elements.
- Select the contractor most suited to the needs of the turnaround.
- Agree achievable, measurable targets with the contractor – on safety, duration, hours expended, quality, or any other useful indicator.
- Ensure the contractors' IT systems are compatible with the client's systems or that there is an effective translation system.
- Ensure that, if the contractor is planning the overhaul work, the plant shutdown–start-up network can be linked to the contractor's schedule.
- Place the order for the contract as early as possible, to give the contractor time to prepare for the contract.
- Prior to the start of the event, thoroughly brief the contractor's employees on the standards required.

Factors influencing the selection of contractors

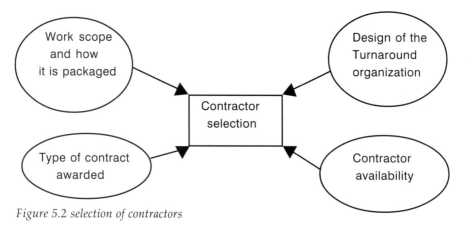

Figure 5.2 selection of contractors

- Ensure that all necessary documentation is issued before the time it is required.
- Ensure that skill levels are as promised by the contractor.

During the event

- Closely monitor the contractor's performance against the agreed targets and ensure steps are taken to rectify any sub-standard or late work.
- Ensure the client organization does nothing to hold the contractor up (the most important matter here is the timely issuing of permits to work).
- Minimize knock-on effects from one contractor to the next by effective co-ordination of the total task.
- Speedily resolve all client–contractor and contractor–contractor conflicts.
- Ensure all variation instructions and emergent work requests are approved and issued in written form.
- Ensure contractors understand that they will not be paid for any work they have carried out on a verbal instruction.
- Validate contractor claims at the earliest opportunity.
- Sign off completed contractor work as early as possible.

After the event

- Thoroughly debrief the contractor to establish 'lessons to be learned'.
- Score contractor's performance and retain for future reference.
- Develop improved contractor programmes for future events.
- Gather and quantify the history of the overhaul (extremely useful data for the planning of future turnarounds).

Contract management is becoming a larger part of turnaround management as companies rationalize, divest and downsize. The emphasis is therefore moving from managing the work to managing contracts.

Case study

A client contracted a very large and well-known construction company to schedule and execute and overhaul a turnaround. The client had already planned out the majority of individual jobs on method sheets using Lotus Approach. Before the contract was signed it was assumed (by both client and contractor) that the large data base of information could simply be downloaded from Lotus Approach to the contractor's PrimaVera scheduling software in order to produce the schedule for the turnaround. In the event, the IT representatives from both companies found it impossible to do this and the result was that after much agonizing and negotiation between client and contractor, it was agreed that the latter would take on extra staff (at the client's expense) to manually key in the data which was read from hard copies of the Lotus Approach reports.

The client checked to ascertain if any difficulties would be experienced in linking the plant shutdown and start-up networks to the contractor's schedule. It was discovered that, for this task, the client had used PrimVera software and it was agreed this would be compatible with the contractor's software – also PrimaVera.

Unfortunately, when the time came to link the networks to the schedule it was found that, although both programmes were on PrimaVera, the client had a later version of the software and the link up was never made.

Once again the data had to be input manually. This meant even more staff had to be put on planning (again at the client's expense). A far greater consequence was that the turnaround schedule was not ready for the start of the event. The critical path, and therefore the duration of the event, was unknown. From day one the turnaround was out of the manager's control.

What could have been done to avoid this?

6
Planning the turnaround

Introduction

A turnaround plan is formulated over a long period of time in preparation for a large amount of work which is going to be performed in a very short space of time. It is not uncommon to plan for nine months a turnaround which will be completed, product to product, in less than three weeks. The normal turnaround features a high volume of work carried out by a large number of people working under extreme time and access constraints. It therefore requires planning of an order and detail that is not found during normal production.

The basic objective of planning is to ensure that the right job is done at the right time by the right people. Figure 6.1 shows how this objective is met using work specification, work scheduling and resource scheduling.

The planning of a turnaround requires the active participation of a large number of people including, among others, the following:

1. *Preparation team*: under the leadership of the turnaround manager create the plan which will be presented to the policy team.

Planning strives to ensure that the right job is done at the right time by the right people, therefore the main activities are:

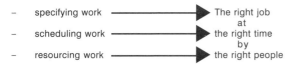

The planning of a turnaround requires the active co-operation of many people, including:

-   Preparation team
-   Plant personnel
-   Inspectors
-   Engineering department
-   Project managers and engineers
-   Contractor's representatives
-   Suppliers' and vendors' agents
-   Equipment manufacturers

Production of the master plan is the responsibility of the turnaround manager and preparation team

Figure 6.1 The basic objective

2.   *Plant team*: provide basic data, work requests, technical information and the shutdown–start-up network and then validate final planning documentation.
3.   *Inspectors*: specify inspection work, requirements and techniques.
4.   *Engineering department*: provide technical information and support on a range of topics.
5.   *Project managers and engineers*: provide the planning and documentation for project work.
6.   *Contractor representatives*: advise on feasibility of their part of the plan.
7.   *Policy team*: approve and fund the final plan.

The main elements in creating a turnaround plan are shown in Figure 6.2 and demonstrate the necessity for co-operation from all those concerned.

The first stage in planning is the creation (at a series of 'Work List Review Meetings') of task specifications, considered under three basic headings – major tasks, small tasks and bulk work – a different planning method being adopted in each case.

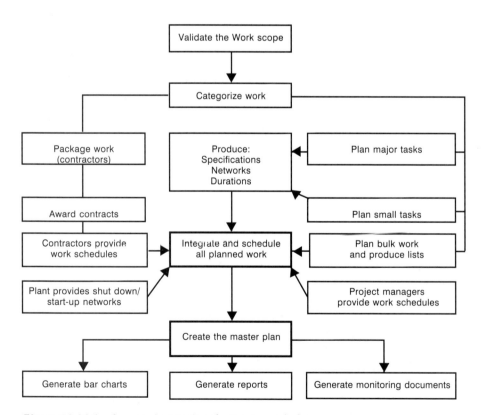

Figure 6.2 Main elements in creating the turnaround plan

Major tasks (see Figure 6.3)

A major task is one that fulfils one or more of the following criteria:

- it is abnormally hazardous;
- it is technically complex or unfamiliar;
- it has a high work content;
- it will involve a large number of people with different skills;
- it involves multi-level working;
- it will take a long time to complete.

Check list of requirements

1. Major task method statement
A detailed document which defines the following:

- step-by-step procedure for performing the task
- list of fully specified materials
- day-by-day profile of manpower requirements
- day-by-day schedule of tools and equipment
- day-by-day schedule for services requirements
- day-by-day schedule of utilities requirements
- notification of any special circumstances or requirements

2. Task networks
A flowchart which shows the following:

- each task activity in logical sequence
- a duration for each activity
- the 'longest time' critical path
- any allowance for contingency
- slack time for shorter activities

3. Isolation register
A document which contains:

- a sketch or list defining the position of every isolation required to make the task safe
- the type of isolation at each position, e.g., slip plate, spectacle plate, physical disconnection
- a detailed schedule of the isolation plate ratings
- a certificate which the plant responsible person must sign to indicate isolation is satisfactory

4. Welding procedure
A document which contains:

- a sketch of the butt to be welded
- parent metal and electrode material specifications
- weld process to be used
- pre and post weld heating requirements
- rectifier type and weld current
- weld test required to qualify welders
- none destructive test requirements

5. NDT request and report
A document which details:

- the type of non-destructive test required
- the date and time the test is required
- a report sheet for detailing the results of the test

6. Pressure test request and report
A document which details:

- plant item number/line number
- test medium and temperature
- test pressure and duration
- sketch of feed and bleed points
- type of equipment and pressure gauge number
- pressure and temperature recorder numbers
- pump details
- name of tester and acceptor
- results of test, and comments

7. Task safety plan
A document containing:

- a task hazard report
- a rescue plan

8. Any other necessary documents
Any documents needed to clarify or assist the performance of the task, e.g.:

- statutory documents
- assembly or detail drawings
- standards or specifications
- plant SOPs
- manufacturer's or vendor's recommendations

Figure 6.3 Major tasks

Some examples of major tasks are:

- overhauling a large machine;
- replacing all of the elements of a long conveyor belt;
- applying an internal protective coating to a large vessel with a small manway access;
- demolishing a redundant cooling tower;
- installing a new computerized instrumentation control system (such as a DCS);
- re-traying a large distillation column.

Such tasks normally demand the input of an engineer. In the case of turnarounds it is the preparations engineer who works up the work packages for the major tasks. His job is to transform the basic data into a set of written instructions and supporting documentation. This must provide sufficient information to enable someone to carry out the task with as little reference to outside sources as possible – which means that the major task work-up must cover *every* known aspect of the task (see Chapter 3, where the major task review meeting was discussed in detail).

The following information package is created by the review team led by the preparations engineer (see Figure 6.3 for a detailed listing of the requirements of the information package):

1. Statement of major task method.
2. Task network.
3. Isolation register.
4. Welding procedure (if required).
5. NDT request and report forms.
6. Pressure test procedure and report forms (if required).
7. Safety plan (see Chapter 10).
8. Any other necessary documents.

When the task package is complete the preparations engineer passes it to the planning officer for inclusion in the turnaround schedule. At a later date, usually two to three weeks before the start of the event, the preparations engineer will present the information from a representative sample of the major task packages to the policy team, and other interested personnel, at the final review meeting for major tasks.

Minor tasks (see Figure 6.4)

Minor or small tasks do not require the input of an engineer but they do require individual specification by an experienced planner. Typical such tasks are:

- clean, inspect and if necessary repair a medium size or large heat exchanger;
- inspect and if necessary repair a small vessel;
- replace a nest of twenty-four steam traps;

Check list of requirements

1. Small task method sheet

A standard form which specifies the following:

- Order number and task number
- Tag number and location of task
- The job method
- A materials list
- Manpower required
- Tools and equipment required
- Services required
- Utilities required
- Any special requirements

2. Small task networks

A simple flowchart cross-referenced to the task method sheet, showing the following:

- The main steps of the task in logical sequence
- A duration for each main step
- An overall duration
- Any contingencies

3. Isolation register

A standard form, cross-referenced to the task method sheet, specifying:

- A sketch or list defining the position of every isolation needed to make the task safe
- The type of isolation at each position, e.g., slip plate, spectacle plate, physical disconnection
- A detailed list of isolation plate ratings
- A certificate which a responsible plant person must sign to indicate isolation is satisfactory

4. Welding procedure

A standard form, cross-referenced to the task method sheet, specifying:

- A sketch of the butt to be welded
- Parent metal and electrode specifications
- Weld process to be used
- Pre and post weld heating requirements
- Welding rectifier type and weld current
- Weld test required to qualify welders
- None destructive test requirements

5. NDT request and report

A standard form, cross-referenced to the task method sheet, specifying:

- The type of NDT task required
- The date and time the test is required
- A report sheet detailing the results of the test

6. Pressure test request and report

A standard form cross-referenced to the task method sheet, specifying:

- Plant item number/line number
- Test medium and temperature
- Test pressure and duration of test
- Sketch of feed and bleed points
- Type of equipment and pressure gauge number
- Pressure and temperature recorder numbers
- Pump details
- Name of tester and acceptor
- Results of test, and comments

Other considerations

Figure 6.4 Small tasks

- remove, overhaul and replace a large pump.

The planner produces, as necessary, a combination of some or all of the following documentation (Figure 6.4 gives a more detailed listing of the requirements of the information package):

- Task method sheet;
- Small task network;
- Isolation register;
- Welding procedure;
- NDT request and report form;
- Pressure test procedure and report form.

Once the planner has completed the small task package he passes a copy of it to the planning officer for inclusion in the turnaround schedule. He retains the original that will later be bound – with other relevant task packages – into an area, unit or work-type planning book for issue to the supervisor who will execute the work.

Bulk work (see Figures 6.5 and 6.6)

Bulk work consists of large groups of identical, or similar, simple tasks that do not need to be specified in individual task sheets but can be bulked together on one list. The planner produces the list of work, typically covering the overhaul or replacement of such items as:

- valves (control, relief, isolation, return etc.);
- small pumps;
- bursting discs;
- orifice plates;
- other simple units.

Network plans are not needed for these jobs because they are normally included in the schedule as blocks of work. This is done purely to derive a meaningful resource profile, because existing computerized schedulers are not ideal for handling large numbers of short duration tasks. The scheduler will use the short duration tasks individually to fill holes in the resource requirement profile caused by the quiet periods on major jobs. Consider, for example, a relief valve and a control valve adjacent to each other on top of a thirty metre high column. To fill resource gaps, the schedule may programme removal of each valve on a different day. This is clearly a waste of time and resource. Therefore, valves and other small items are normally scheduled and marshalled manually. Figure 6.7 shows a manual valve marshalling sheet which defines the number of valves to be removed each day and the order in which they are to be decontaminated, sent for overhaul and returned to site.

Thus far, it has been assumed that tasks fall neatly into one of the three categories discussed, viz major, small, and bulk work. What is to be done,

YEAR............ PLANT................ UNIT / AREA.............. ITEMS............... NO: ....

PLANT ITEM NUMBER	OVERHAUL OR FIT SPARE?	ASSOCIATED PLANT ITEM	TAG NUMBER	GRID REFERENCE	HEIGHT FROM GROUND IN METRES	WEIGHT OF ITEM IN KG	SCAFFOLDING REQUIRED	DELAGGING REQUIRED (ASBESTOS ?)	CRANE SIZE REQUIRED	MECH.	ELEC.	INSTR.	CIVIL

Figure 6.5 Bulk work specification sheet

YEAR.......... PLANT.................. UNIT / AREA.......... ITEMS.......... No: ......

ITEM NUMBER	ASSOCIATED PLANT NUMBER	INST / ELEC DISCONNECTION (IF REQUIRED)	REMOVED	DECONTAM.	SENT TO WORKSHOP FOR OVERHAUL	RETURNED FROM WORKSHOP	OVERHAULED ITEM FITTED	SPARE ITEM FITTED	INST / ELEC RECONNECTED (IF REQUIRED)	TESTED AND COMMISSIONED

Figure 6.6 Bulk work control sheet

Dispatch to workshops					Latest acceptable return from workshops					
DAY 2	DAY 3	DAY 4	DAY 5		DAY 23	DAY 24	DAY 25	DAY 26	DAY 27	DAY 28
CV1001	CV1012	CV1023	CV1234		CVIOOI	CV1102	CV1013	CV1023	CV1234	RV132
CV1O22	CV1013	CV1006	RV132		CVIO22	CV1209	CV1014	CV1006	IV1108	RV133
CV11O2	CV1014	CV1007	RV133		RVIO5	CV1012	VR108	CV1007	IV1109	IV1200
CV1209	RV108	CV12O4			RV106	RV101	RV110	CV1204		
RV105	RV110	CV1205			IV1006	RV109	RV102	RV123		
RV1O1	RV102	RV122			IV1007	RV189	RV122	RV124		
RV109	IV1012	RV123					IV1012	RV125		
RV189	IV1008	RV124					IV1008	IV1102		
IV1006	1V1009	RV125					IV1009			
IV1007		IV1102								
		IV1108								
		IV1109								
		IV1200								

Sequence determined by the needs of the schedule and by accessibility of the items

Figure 6.7 Valve marshalling sheet

however, with a task that falls on the border line between two categories? Remember that a turnaround is a hazardous event – take no risks! If there is any doubt whatsoever, always elevate the task to the higher category.

The list of tasks provides the raw data for defining what is sometimes referred to as the 'mechanical duration' of the turnaround, i.e. all tasks that are carried out between the time the plant (or any relevant section of it) is shut down and the time it is started up. The additional information that is needed to produce a schedule for the total event is termed the 'shutdown–start-up logic'.

The shutdown–start-up logic

Options (see Figures 6.8 and 6.9)

The combining of the shutdown–start-up logic with the mechanical duration logic will have a fundamental effect on the planning and execution of the turnaround. There are four main options and each will involve the turnaround manager in a different role with different responsibilities. The options are

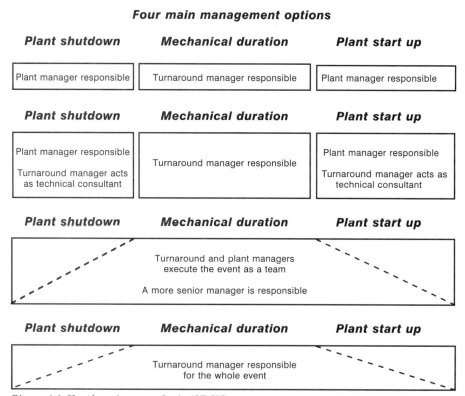

Figure 6.8 Shutdown/start-up logic (SDSU)

1. Mechanical duration only

- *Features (Separate steps)*
- Plant staff shut the plant down, decontaminate and cool it then hand it over to turnaround team
- Turnaround team complete all planned work and hand the plant back to plant staff
- Plant staff start the plant up assisted by a start-up team supplied by the turnaround manager
- Well defined cut-off points
- Plant manager is responsible for shutdown/start up
- Turnaround manager is responsible only for the mechanical duration
- *Benefits*
- Turnaround planning simpler and cheaper
- No formal shut down/start-up network required
- Plant may be shut down/started up in any order
- Common start for manpower
- Overrun costs charged to the plant
- Risk of accidents/incidents less than Options 3 and 4 but greater than Option 2
- *Drawbacks*
- Product to product duration will almost certainly be longer
- Production loss will almost certainly be higher

These drawbacks may not apply if the critical path task has the longest shutdown/start-up duration

2. Mechanical duration + consultancy

- *Features (Separate steps)*
- Turnaround team advise on best practice for shut down/start up programmes
- Plant staff shut the plant down, decontaminate and cool it then hand it over to turnaround team
- Turnaround team complete all planned work and hand the plant back to plant staff
- Plant staff start the plant up assisted by a start up team supplied by the turnaround manager
- Well defined cut-off points
- Plant manager is responsible for shut down/start up
- Turnaround manager is responsible only for the mechanical duration
- *Benefits*
- Duration may be shorter than in Option 1
- Production loss may be less than in Option 1
- Custom and practice is challenged to find more effective ways of shutting down and starting up
- SDSU overrun costs charged to the plant
- Less risk of accidents/incidents than other options
- *Drawbacks*
- Turnaround team workload is increased
- A formal shutdown/start-up network is required
- If network is not adhered to, work may be delayed

3. Integrated plant/turnaround team

- *Features*
- Plant/turnaround staff form one team and produce an integrated turnaround plan, product to product
- The plant is shut down, decontaminated and cooled system by system, in a pre-planned sequence
- Equipment is opened up and worked on as soon as safe, while other systems are being shut down
- Systems brought on line in a planned sequence while other systems are still being overhauled
- Turnaround and plant manager manage jointly

A senior manager has overall responsibility
- *Benefits*
- Product to product duration can be significantly shorter and product loss less than Options 1 and 2
- A mixture of plant team local knowledge and the turnaround team expertise ensures increased effectiveness and better practice
- *Drawbacks*
- Planning is more complex and costlier
- Integrated product to product network required
- If network isn't adhered to, work may be delayed
- More complex phased start dates for manpower
- Greater risk of accidents/incidents than in other options

4. Total event

- *Features*
- Turnaround manager is responsible for the total event, product to product
- Turnaround team has its own operations and process personnel
- The plant team hand the live plant over to the turnaround team who shut it down, overhaul it, start it up and hand it back to the plant team
- *Benefits*
- Greatly reduced interface problems
- Product to product duration can be significantly shorter and production loss less than other Options
- Well practised team in total control can greatly increase effectiveness even over Option 3
- *Drawbacks*
- Planning more complex and costlier
- Product to product plan required
- Lack of local knowledge may be a problem
- Overrun costs borne by turnaround team
- Risk of accidents/incidents greater than Options 1 and 2 but less than Option 3

Figure 6.9 Characteristics of the four SDSU options

listed below (also see Figure 6.8) and the features, benefits and drawbacks of each are defined in Figure 6.9.

1. The turnaround manager is responsible for the mechanical duration only and the plant manager for shutdown and start up.
2. The turnaround manager is responsible for the mechanical duration, the plant manager for shutdown and start up with turnaround manager acting as a shutdown–start-up consultant.
3. The turnaround and plant managers combine as a team for the total event with a more senior manager having overall responsibility.
4. The turnaround manager is responsible for the total event.

The option chosen will depend upon the culture of the client's company and the constraints operating at the time, which (among other things) could include:

- plant team's expertise (How well would they know their plant?);
- turnaround team's expertise (Especially the manager's);
- amount of money available (To buy in expertise);
- amount of time available (Would it permit a particular option to be implemented?);
- plant profitability (If high, would necessitate a quick turnaround);
- plant utilization (May be low, so a lengthy low-cost turnaround may be acceptable);
- plant complexity (May force the choice of a particular option).

The constraints have to be balanced and the options carefully considered. It cannot be stressed enough that this is one of the most crucial decisions that will be made during the preparation phase. It is worth repeating that it will have a fundamental effect on the planning and execution of the event. For that reason, the decision on which option to adopt must be taken by the policy team.

Having described the options available for combining the shutdown and start-up logic with the mechanical duration, it becomes possible to define the shutdown and start-up networks.

The shutdown network

The shutdown network is a critical path programme and its supporting documents, which define (in logical sequence) the activities required to bring the plant off line and prepare it for handover to the turnaround team.

The critical path is created by the plant team and should comprise the following elements:

- the shutdown logic for each plant system;
- identity of significant equipment within each system;
- the order in which each system will be shut down;
- the individual activities performed in the control room at

each stage;
- the individual activities performed on the plant at each stage;
- whether activities are concurrent or sequential and how they are integrated;
- the duration of each activity;
- waiting time (if any) between specific activities;
- total duration for shutting down each system;
- total duration for the complete shutdown.

Other information required includes:

- temperature, pressure and flow-rate of fluids to be used;
- specification of any chemicals to be used;
- equipment, services and utilities required;
- numbers and types of manpower resources required;
- safety plan for the shutdown.

As with the planning of major tasks, the formulation of shutdown networks must take account of every known aspect of the shutdown. If it is intended that mechanical work will be permitted on some systems while others are still in the process of shutting down then the earliest start times for mechanical work must be indicated.

Start-up network

The start-up network is similar in format to the shutdown network and all of the elements listed above for the shutdown must be specified in the process of defining the start-up sequence. An additional requirement during start-up will be the testing of all relevant components, a programme which could include instrumentation tests, pressure tests, relief valve tests, product quality tests etc. – all of which must be written into the logic of the network.

The critical path programme (see Figure 6.10)

When the logic has been worked out for shutdown and start-up, a bar chart is constructed showing the total shutdown and total start-up sequences and durations. During scheduling, the mechanical durations (especially those of the critical path tasks) are inserted between the shutdown and start-up elements so that a 'product to product' duration can be calculated, by summing lag time, shutdown duration, mechanical duration and start-up duration. The critical path task is the task which is completed last, and is not necessarily the longest duration task (an important point which is sometimes missed). The 19.75-day job which cannot be started until Day 4 will be the critical path, not the 22-day job which can be started on Day 1.

Work scheduling (see Figure 6.11)

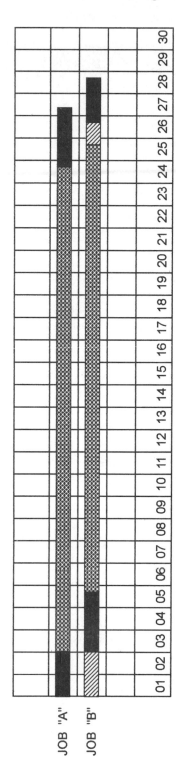

1. Job A has a mechanical duration of 22 days, Job B mechanical duration is 19.75 days and based on this alone, Job A would be considered the critical path job.

2. If the shutdown and start up time for each job was known Job A would be 26.2 days long and Job B would be 25 days long - Job A would still be considered the critical path job.

3. However, it is only when the logical relationship between the two Jobs is known that the true critical path can be calculated.

4. Due to the lag time at the beginning (Job B cannot be shut down until Job A is off line) and at the end of the mechanical duration Job B is the longer job by 1.5 days and is therefore the critical path.

4 The critical path is calculated from the first activity in the shutdown until the last item of equipment is back on line, and that is not necessarily the 'longest' job.

LAG TIME　　SHUTDOWN AND START UP DURATION　　MECHANICAL DURATION

Figure 6.10 Critical paths

Main elements required

Constraints	Schedule

Constraints		Schedule
– fixed duration – resources – working patterns – budget	▶	1. Input initial duration 2. Initiate logic register 3. Input shutdown networks
Shutdown/start-up programmes	▶	4. Input start-up networks
Project work programmes	▶	5. Input work packages 6. Input initial working patterns
Contractor's work programmes	▶	7. Run and re-run the schedule – fix duration – smooth out resource profile – manipulate working patterns – manipulate work sequence
In-house planned major tasks	▶	
In-house planned small tasks	▶	8. Using data generated, produce an initial proposal for the following: – work plan – budget – duration – resource profile – working patterns
In-house planned bulk work	▶	
Turnaround generated work	▶	9. Submit the above proposals to the policy team for approval

Figure 6.11 Work scheduling

Options

Depending upon the size of the event, the technology available, and the company's resources of skill, time, money and equipment, a work schedule can be constructed in several different ways, viz:

1.   Using a software scheduling package on the client's mainframe computer.
2.   Using a software scheduling package on a stand-alone PC.
3.   Manually, using a project planning sheet.
4.   Manually, using a 'shuffle board' (a slotted board – similar to those used to hold 'clocking in' cards – which lays out the different elements of task bars, using different coloured cards, to form a complete work schedule. Its advantage over the project planning sheet is that, when circumstances or priorities change, the elements can be

'shuffled' backwards or forwards to quickly represent the new situation.

Objectives

The objective of scheduling is to produce an integrated work programme (which might cover a period of anywhere between one and eight weeks) which will provide:

1. A plan for executing all tasks in a logical sequence.
2. A cost profile which will not exceed the budgetary profile.
3. A duration within the available time scale.
4. A manpower-needs profile which can be resourced.
5. Economical and sustainable work patterns.

The final turnaround schedule will be an optimized blend of the above requirements. The inputs required to build the schedule and the constraints within which it must operate are defined in Figure 6.11

The constraints

The art of scheduling lies in the balancing of constraints. Consider this: if – in a particular case – time, money and resources were unlimited there would be no need to plan or schedule work. It is constraints which define the planner's role.

There are two useful observations to make about constraints. The first is that they are intimately connected to one another. Take four of the basic constraints surrounding a turnaround, viz. duration, cost, resources and working patterns. Each has its own individual impact on the work scope and is, in turn, influenced by the work scope, but each also affects the others. If one changes, the others will either change automatically or will have to be changed to compensate. For example, one way of halving the duration of a turnaround which had been planned on the assumption of single-shift working would be to change to double-shift working – but that would also mean that twice as many people would be needed.

The second observation is that 'constraints' constrain objectives, and objectives are often in conflict. One objective may be to complete the event in ten days. Another may be to use only single-shift working patterns to minimize the number of resources used and keep costs down, but the duration of the longest task in the work scope is planned at twelve days. Clearly, given the current situation, both objectives cannot be met. The constraints have to be balanced. This is the task facing firstly the turnaround manager and then the policy team. It is their job to optimize the turnaround plan.

Optimizing the turnaround plan

Presenting the plan

Once all elements of the plan have been finalized by the turnaround team, the turnaround manager presents them to the policy team for final discussion and, it is hoped, for approval. Should the team reject the plan at this stage (and they quite often do) their reasons for doing so might well include one or more of the following:

- the duration is longer than they desire it to be;
- the cost is too high;
- the resources are greater than the perceived plant saturation level;
- the working patterns are undesirable.

Optimizing the plan

In this case, the policy team must balance the constraints imposed upon the turnaround. There are ten main options for balancing constraints. The first eight are:

(1) *Increase the duration of the event* – so that the amount of work required can be accomplished with less people or a less intensive shift pattern.
(2) *Reduce the duration* – if it is thought to be too long. This will mean either increasing resources or altering the shift pattern. However, if the duration of the event is determined by that of the critical path task, it will not be possible to do so – unless the critical path task is eliminated.
(3) *Increase the cost* – in order to reduce one of the other constraints. Again, this is not always possible.
(4) *Reduce the cost* – this will mean reducing something else also.
(5) *Increase manpower* – in order to reduce duration but, depending upon available resources, this may not be possible.
(6) *Reduce manpower* – to reduce costs.
(7) *Increase the use of shift work and overtime* – to reduce duration. Once again, this may not be possible.
(8) *Reduce shift work and overtime* – to reduce costs.

These are conflicting options. The policy team may exercise any combination of them to achieve an 'optimized' plan. How they are exercised will depend largely on which constraints the team perceive to be the critical ones. Time? Money? Resources?

The ninth option is to reduce the work scope. Remember this is at the point where all tasks on the work list have been technically justified. If this option is chosen it is for business reasons not technical ones. In this sense, technically, it could be said that the policy team are taking a risk. They will have to bear the consequences if the work omitted results in an adverse effect on plant reliability at a later date.

The tenth option is more involved, much more challenging, but potentially very rewarding. It involves the challenging of current assumptions about how work is done. It is used primarily in the attempt to reduce duration but obviously has an effect on all other constraints. The turnaround manager forms a team of those people he considers to be best suited to the task at hand. Their purpose is to challenge, using creative thinking techniques ('brainstorming'), every element of the longest duration jobs on the work list in order to reduce the time scale.

A long duration job consists of a sequence of different activities carried out by different people. The activities have been planned in a particular sequence, using specific methodology, technology and resources. Most of all, the plan is the visible result of a sequence of decisions made on the assumptions of an individual or a small group. The methodology or the technology, for instance, may have been chosen purely because 'that's the way it has always been done'. The creative session is organized to break through those types of assumption to a newer, more effective way of doing things.

What has to be asked is whether time can be saved by:

- eliminating the task altogether (alternatives? consequences?);
- altering the content of the task;
- altering the sequence of activities;
- performing sequential activities in parallel;
- reducing the duration of, or eliminating, any individual activity;
- reducing or eliminating gaps between critical path activities;
- altering either the number or the mix of manpower;
- altering the shift pattern on this particular task;
- using a different methodology;
- using a different technology.

Reality teaches that there are many ways of achieving the same end, assumptions limit us to only one way.

Contingency planning

The challenge

As discussed in chapter one, turnarounds deal with plant that may be old, worn out, damaged or malfunctioning. For this reason, contingency planning is probably the most vexed and emotional element in turnaround planning. So much so that it is often ignored – and the omission later regretted. The reason for this is that, in the context of a turnaround, contingency planning is an activity which builds extra time, money and resources into a plan to cover for emergent work – tempered by the knowledge that it might never happen. This uncertainty leads to much debate about the need, and in extreme cases the ethics, of building preparation for contingency into a plan. On one hand if the turnaround manager builds in what is considered to be too much contingency allowance he may be accused of 'padding out' or 'building fat into' the plan for his own comfort. He will get little support from the policy team especially if, in the event, the magnitude of emergent work is less than predicted. On the other hand, if the turnaround manager builds in no preparation for contingency, or too little, and a significant amount of such work does emerge, causing an over-run on time and costs, he will carry the responsibility for the business consequences of the over-run. As an example, take the task request for a particular vessel, which reads 'Inspect and if necessary repair...'. How should this be treated? The person who wrote the request has raised the possibility that the internals may be damaged or worn, and may need repairs (which may be a significant one) but, when questioned, he states that he does not know for certain whether a repair will be needed but believes that it might. He has established his concern and it is now up to the turnaround manager to act.

Does the turnaround manager build in time and resources for such contingencies – and, if so, how much? He can consult with technical experts but they can only advise on how long a certain repair might take (all other things being equal) or how much it might cost (providing everything goes to plan). They cannot answer the two vital question he needs answered. Namely, will a repair be required? And if so, what will be the extent of it? They just don't know – nobody does – nobody can. The item in question is hidden from view. The turnaround manager must therefore guess – nothing more scientific than that. The guess will be based on his (and other's) experience of similar past events, so-called 'engineering judgement' – but still a guess!

The one sure thing is that there will be an amount of emergent work. An informal survey of twenty turnarounds performed on older plants showed a low of 5 per cent and a high of 47 per cent cost over-run due to emergent work. These figures are indicative only but a spread as wide as this, with a mean of 26 per cent, could lead to the 'too much – too little' situation described above.

In one actual case a large vessel had a suspected crack in the shell. A contingency allowance of two days was built into the plan for weld repair, heat treatment and radiography – it was agreed that this was a tight target – no one new just how tight. In the event, the crack was found but it was almost three metres long and fifteen millimetres deep. The repair was completed twelve days later.

A second case involved a packed column in which it was suspected that there was something wrong with the table that supported the packing. On inspection it was found that the table had capsized and was blocking the hole from which the packing was normally extracted to allow access to the table. One day's contingency allowance had been made for any necessary repair. The normal one-day activity of removing the packing took five days. The table was found to be undamaged but the time had already been lost.

As far as planning for contingencies goes, The turnaround manager can only exercise his best judgement and get the commitment of the policy team to support the allowance calculated or take joint responsibility for altering it. Another practice is to build in a general contingency reserve for the whole event, but this too is only a guess and prone to all of the vagaries explained above.

Defining emergent work

There are two types of emergent work: extra and additional.

>   *Extra work* – is generated by an existing task, as in the above examples of inspecting an item and finding a fault which must be rectified. It can also come about as a result of damage inflicted on the item by the team which is performing the task or by someone else.
>
>   *Additional work* – is tasks added to the work list during the event. This may happen because the plant team simply forgot to include them on the approved work list (even after all those work list meetings!), or because of something that happens during the event (this is most likely to occur when the plant is being brought back on line).
>
>   One common fault is that pumps – which may not have been included in the turnaround – fail on start-up and must be replaced or repaired. Unfortunately at that late date there is no opportunity to recover the time lost.

Both types of work must be closely controlled (see Chapter 13).

Several iterations of the process of optimizing the plan are likely to be needed before it is finally approved by the policy team. It should then be widely published so that everyone concerned with the event is able to peruse it and, if necessary, challenge any of its elements. Once key personnel have agreed to the plan, all subsequent activities must conform to it. The only exception being that if the situation changes, or an element of the plan is found to be ineffective, the policy team will reconvene to suitably adjust the plan. Using the approved plan the turnaround team then generate a number of documents to allow them to monitor and control the flow of work during the event.

Generating control documents

The purpose of generating control documents is to:

- Quantify available resources (How much? What skills?)
- Define work to be done on a daily basis.
- Track performance and work achieved.
- Flag up if the event is ahead of, on or behind target.
- Indicate future progress.
- Indicate need for corrective action.
- Maximize use of vital resources.
- Minimize expenditure.
- Indicate priority areas.
- Indicate the impact of unpredicted problems.
- Help the turnaround manager to stay sane.

Typical documents generated are:

Reports	– Display current achievements, e.g. 'S' curves, bulk work control sheets and schedule updates;
Worksheets	– Define workload, e.g. daily bar charts and punch lists;
Work control sheets	– The visible aspect of single point responsibility; Each step in a task is 'signed off' by the person responsible for it. The next step is not started until the signature is in place.

Having put together the turnaround plan, the turnaround manager needs to build an organization to execute it, a task which will be dealt with in the next chapter.

Case study

This case study will highlight that it is human beings who plan turnarounds, not computers. While this may seem obvious to some, to others it may come as a surprise.

A particular turnaround had been cancelled by the policy group, on the mistaken assumption that this would have no adverse effects on related plants. Six weeks before the programmed due date it was discovered that the turnaround would indeed have to be done, and on the original date because it interacted with the operation of another larger plant.

A policy team meeting was hastily convened and the turnaround manager was asked to put together a plan for the work (some 270 tasks, six of them major ones). There was no time to plan the event on computer, generate a schedule or produce the normal task sheets. A scratch team was assembled and – after the initial shock and denial stage was overcome – they agreed to put aside their previous training (which placed the computer at the heart of the planning and scheduling effort) and plan the event manually. The schedule was created by cutting, pasting and colour-coding sections of the work list (Monday's work coded yellow, Tuesday's green, and so on). This exercise also generated the critical path and duration. Major task instructions were written down on A4 sheets and the supervisors who would do the jobs closely briefed by the turnaround manager. The remainder of the work was put on punch-list control sheets and the only detailed work which was carried out was the specification of materials.

The event was completed within duration and cost. At the post-event de-briefing, members of the team stated that, although it had been extremely hard work, they had felt more in control of the event because (a) they had personally contributed more to the logic of the plan during the preparation phase, and (b) the fact that they did not have a computerized schedule to update daily meant that they spent a lot more time on-site, monitoring, advising and assisting.

Admittedly this was a relatively small event but it did bring home the point that the intelligence in planning and scheduling resides in the human beings and not in the computers. Our increasing dependence on computer systems – hugely effective as they are – should not blind us to this fact.

7
The turnaround organization

Introduction

The turnaround organization – the blend of people who will execute the work on the day – is critical to the success of the event. Management must think thoroughly through the issues involved, select the most suitable people available at the time, and organize them with great care in order to forge the strongest possible organization for controlling the turnaround. The organization is the template which will shape the course of the turnaround and any flaw or weakness in it will infect every other aspect of the event.

Organizational combinations (see Figure 7.1)

The shape and size of the organization will be determined by separately addressing two key questions, i.e.:

(1) Who will manage the turnaround?
(2) Who will carry out the work?

Taken together (as they must be), the answers to these questions will reveal that a large variety of organizations is possible.

Who will manage the turnaround?

As explained in Chapter 2, there are three options available when choosing a turnaround manager, the choice being based on a number of considerations.

Option M1 (where M = Manager)

Select one of the company's own management or engineering staff. The considerations here will be ones of availability and competence. A kind of half-way house between this option and M2 is to select a member of staff to manage the turnaround and then bring in an experienced consultant to 'advise' him.

Option M2

Bring in a consultant management team. More and more companies offer this service. The considerations are price, (they can be expensive) and company culture (do we really want to put our future into the hands of strangers?).

Option M3

Bring in a main contractor to manage and execute the bulk of the work. The consideration here is one of trust. Whereas the consultant management team may be strangers they are at least independent of the contractors who will execute the work. On the other hand the contractor manager has a vested interest which might lead him to 'milk' the contract to maximize profit or take decisions which are favourable to his company and may be to the disadvantage of the client. Again, company culture will determine if this is a viable option.

Once the manager (or management team) is chosen, the next question is:

Who will do the work?

There are six main options to consider;

Option W1 (Where W = work organization)

Employ one main contractor who will execute all work packages. World-wide, several companies claim to be able to handle the total package of work. At present (1999), their quality is variable.

Option W2

Employ one main contractor who employs, and manages, sub-contractors for a number of packages. This has the advantage that the turnaround manager oversees only the interface between the control team and the contractor. The main disadvantage is that it introduces another layer of management and can complicate communications.

Option W3

Employ one main contractor, executing general tasks, plus several specialist contractors. This is similar to Option W2, but in this case the turnaround manager interfaces directly with the main contractor and the specialist contractors. The advantage is a flatter organization. The disadvantage is that the turnaround manager has many more people to deal with.

Option W4

Employ several contractors executing specific packages of work. Again this is similar to option W3, but there is no main contractor doing the bulk of the general work. The advantage is that the turnaround manager has an even tighter interface with each contractor. The disadvantage is that because a main contractor would normally handle the semi-skilled and unskilled work within his contract, his absence means even more contracting companies will need to be used, further increasing the number of people the turnaround manager has to deal with.

Option W5

Rely on contract agencies only to provide manpower, the company supervising it, and all other resources, itself. From a client's perspective the advantages are that he directly controls the labour and therefore the execution of the work, and does not have to pay all the contractor management fees. The main disadvantages of using agency labour were discussed in Chapter 5 ('Using contractors – the downside'). There is a further one, however; supervising agency labour can be very difficult, especially when undertaken by plant-based supervisors who are used to leading a small efficient team that (to a great degree) co-operates with the supervisor. Contrast this with the large amount of agency labour needed to perform a turnaround, and consider also the following observations:

- Agency workers will be strangers to the plant, and are only interested in the turnaround as a temporary source of income.
- Productivity can be very low, which usually triggers the emotional reaction to 'throw more men at the job', thus complicating matters even further.

Option W6

Use one's own resources and supervision. This is not normally an option for the total package of work. Most companies will have their own staff performing some of the tasks – almost certainly higher voltage electrical work. The advantage in using local knowledge must be weighed against the disadvantages of labour agreements between management and unions regarding shift patterns, overtime and special payments (probably ideal for working on an operating plant) which can be a hindrance to the effective performance of the turnaround.

Clearly, the number of different possible organizational structures that can be created by combining any of the 'M' options with any of the 'W' options is not small (see Figure 7.1).

Basic principles

The policy team and turnaround manager are responsible for the effective execution of the turnaround. They must ensure that all of the elements that go together to make up the event are properly organized. A number of principles have been developed out of past experience to guide them in their endeavours.

There are various options for executing a turnaround

Combining the elements in the boxes below generates over
twenty different organizational combinations

The option chosen determines the contracts awarded

Who will manage the turnaround?
- Client company management
- Consultant management team
- Main contractor management

Who will do the work?
- One main contractor who executes all turnaround packages
- Main contractor who subcontracts some of the packages
- One contractor executing general tasks plus specialist contractors
- A number of contractors executing specific packages of work
- Contract resources and supervision
- Company resources and supervision

Figure 7.1 Organizational combinations

A turnaround is a task-oriented event

The work list, a series of tasks, is the foundation of the turnaround and everything else depends upon it. For this reason, the task is put at the centre of the planning and preparation effort. All other elements are directed towards the safe, timely and cost-effective performance of the task. This approach generates the particular methodology put forward in this book.

The minimum number of people should be used

A turnaround (especially a major event) involves a large number of people. Careful analysis must ensure that only those people who are necessary are employed on the execution of the event. Problems are not solved by 'throwing bodies at them' and it is more preferable to have a smaller number of higher skilled personnel than vice versa.

The principle is based on the following criteria:

- Each plant has a saturation point beyond which the employment of more people will serve only to decrease useful work done rather than increase it.
- Increasing the number of people increases the risk of communication

Organizational level

Responsibilities

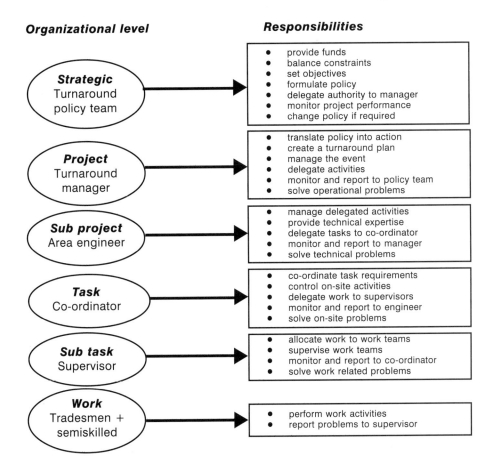

Figure 7.2 The hierarchy

difficulties, and of potential hazards, both to personnel and to their environment.

- The number of highly skilled people is limited, so increasing the number of people on site risks decreases in the overall skill level.
- Costs are increased in proportion to the increase in people.

A turnaround organization is hierarchic (see Figures 7.2

and 7.3)

Much has been written in the last ten years about the limitations of the traditional bureaucratic management structure and the virtues of matrix management and open teams. However, due to the complexity of a turnaround, the organizational model that has, so far, worked

Organizational level	Parent company	Consultant company	Plant personnel	Main contractor	Sub contractors
Strategic level (Turnaround policy team) – provide funds – balance constraints – set objectives – formulate policy – delegate authority – monitor and change policy	Senior engineering, business and financial managers bring their specific expertise and understanding of constraints to the team in order to set smart objectives and create policies to achieve them.	Turnaround manager brings technical expertise in work definition, planning and the management of big events. Acts as the agent of the turnaround policy team.	Plant management provide local knowledge of plant requirements and problems. Responsible for providing the work lists.	Contractor management are only invited to attend policy team meetings in special circumstances and then only to present information and answer questions.	No involvement.
Project level (Turnaround manager) – translate policy – formulate plan – manage the event – delegate activities – monitor and report – change policy	On very large or complex events the policy team may appoint a senior manager to take overall responsibility. The plant and turnaround managers run the event and report to the senior manager.	The consultant turnaround manager possesses skills, specialist knowledge and experience obtained from full time involvement in turnaround management.	The policy team or plant manager may appoint a manager or engineer from the plant to manage the turnaround. Not normally the maintenance manager, to avoid conflict of interest.	The main contractor may offer the option of a total package including the full management and planning team. This should only be considered if the company has a proven track record.	No involvement.
Sub-project level (Area engineer) – manage activities – provide expertise – delegate tasks – monitor and report – solve problems	No direct involvement but may be able to second people from other plants to perform this role. train local engineers.	Consultant companies, when requested, will provide very experienced engineers in key positions to manage areas of turnaround. This strategy is to enhance local performance and	Subject to availability, the plant manager may offer the services of plant engineers to perform this role. Not normally experienced on large events.	The main contractor will normally offer the services of experienced engineers who will work under the turnaround manager. used only as a last resort.	Specialist sub-contractors may provide engineering coverage for large projects. It is not normal for general sub-contractors to offer the service. This option should be used only as a last resort.
Task level (Area co-ordinator) – co-ordinate tasks – control activities – delegate sub-tasks – monitor and report	No direct involvement but may be able to second people from other plants to perform this role.	Will normally provide the planning co-ordinator/officer. Some consultant companies will offer the services of experienced on-site, area co-ordinators.	Subject to availability and need, the plant manager may offer a senior planner as planning co-ordinator. Other trade co-ordinators subject to availability.	The main contractor will normally offer the services of co-ordinators.	Sub-contractor normally offers the services of a general foreman who will act as a co-ordinator. This option should be used only as a last resort.
Sub-task level (Supervisors) – allocate work – supervise work teams – troubleshoot – monitor and report	No direct involvement but may be able to second people from other plants to perform this role.	No direct involvement but can usually recommend manpower agencies which can provide people.	Electrical and instrument supervisors are normally provided by the plant manager. Supervisors for other disciplines are subject to availability and need.	The main contractor will provide supervisors.	Sub-contractor provides supervisors or responds to contractor supervisors.
Work level (Tradesmen and semiskilled) – perform work – report problems	No direct involvement but may be able to second people from other plants to perform this role.	No direct involvement but can usually recommend manpower agencies which can provide people.	Subject to availability, the plant manager may offer maintenance personnel to perform this role.	The main contractor may provide manpower or hire in sub-contractors.	Sub-contractor provides his own manpower.

Figure 7.3 Organizational issues to be considered

best is the hierarchic pyramid. It is effective so long as the responsibilities shown in Figure 7.2 are properly discharged and there is effective communication between levels. The number of levels in the hierarchy will depend upon how large and complex the event is.

Figure 7.3 highlights some of the issues surrounding the selection of key members of the organization. The comments in the various boxes are based on the author's experience and do not purport to be universal. In different situations, in other companies and, indeed, in other countries, different imperatives may apply to the selection of the organization.

Figures 7.4 and 7.5 show models of turnaround organizations. Figure 7.4 models a large event in which the plant was split geographically and the organization was a mixture of plant-based, contract and project personnel. Because it was a large event with a high technical content, it required a large management and technical team (this is what makes performing work on a turnaround more expensive than performing it during normal operations). Note the requirement to have one co-ordinator spanning two areas due to the shortage of competent personnel.

Figure 7.5 models the organization for a much smaller event (consisting mainly of the overhaul of a large number of small pumps and valves) which was manned by plant-based personnel supplemented by some agency semiskilled and unskilled labour. This is typically the type of event chosen for a turnaround manager's first performance in the role, to build up his experience. Although the organization is smaller, note that all of the elements are still represented, albeit in a simplified manner (the 'manager' also carries out the duties of the engineer).

One person must be in overall control

This is a natural consequence of the last criterion. A turnaround is such a compressed event, with much more work being done at higher than normal expense by a large number of people in the shortest possible duration. As a result, there is little time for deliberation or discussion. Many decisions have to be made immediately. To make this possible, one person must be put in overall control of the event.

Single point responsibility is exercised at every stage (see

Figure 7.6)
Much of the work done on a turnaround is of long duration and involves a number of people from different skill backgrounds, departments and even companies. The turnaround, being a process, is defined as 'A sequence of activities performed in logical succession', which prompts the following observations:

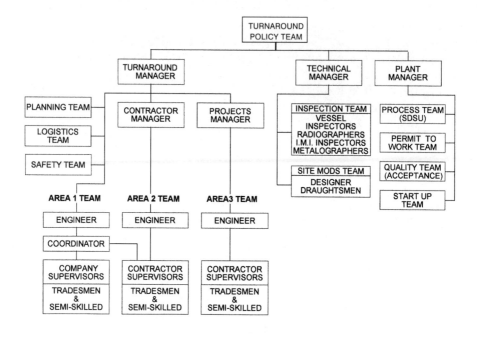

Figure 7.4 A sample Turnaround organization

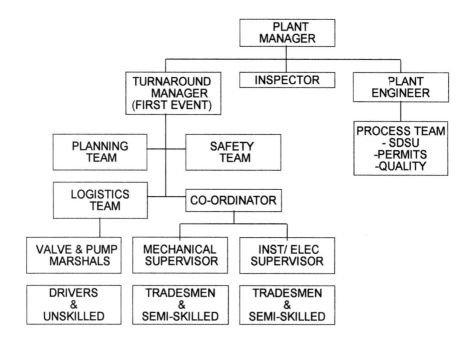

Figure 7.5 Example of an organization for a small Turnaround

Single point responsibility requires that the person in control of the task at any stage has sole responsibility for:

- accepting the task from the previous stage;
- completing the current stage to plan;
- handing the task over to the next stage.

The turnaround manager has sole responsibility for ensuring all stages are linked in an overall plan and everyone is briefed on it.

Work sequence for a small task, i.e.: a vessel with an internal crack

Stage	*Responsible person*
Validate task	Preparation engineer
Plan task	Turnaround planner
Validate plan	Preparations engineer
Schedule task	Planning officer
Validate schedule	Turnaround manager
Obtain resources	Turnaround planner
Brief personnel	Turnaround manager
Shut down plant	Plant manager
Isolate vessel	Maintenance supervisor
Open vessel	Task team leader
Test atmosphere	Safety officer
Install lighting	Electrical supervisor
Install ventilation	Task team leader
Install scaffold	Scaffold supervisor
Inspect/discover defect	Inspector
Troubleshoot problem	Turnaround manager
Create repair procedure	Design engineer
Carry out repair	Welding supervisor
Inspect – NDT	Inspector
Pressure test	Maintenance supervisor
Accept repair	Inspector
Strip out ventilation	Task team leader
Strip out scaffolding	Scaffold supervisor
Strip out lighting	Electrical supervisor
Box up vessel	Task team leader
De-isolate	Maintenance supervisor
Sign completion form	Plant engineer
Trip and alarm test	Instrument supervisor
Start up plant	Plant manager

Figure 7.6 Single point responsibility

- To ensure that the logic of the plan is followed, each activity must be carried out at the correct point in the sequence.
- Each activity must be executed correctly and, if necessary, inspected to verify it.
- There is a gap between each activity. It must be controlled to avoid wasting time.
- If consecutive activities are carried out by different individuals there must be an effective handover.

Throughout the life of the task the responsibility for ensuring that it is performed effectively may change many times – rather like passing the baton in a relay race, so there must be a mechanism to ensure that the baton is not dropped. This mechanism is called single point responsibility and it requires that the persons responsible for each activity must ensure that:

- A good handover is obtained from the person responsible for the previous activity.
- The activity that the person is responsible for is properly executed.
- The task is properly handed over to the person responsible for the next activity.

The turnaround manager is ultimately responsible for ensuring that all tasks are carried out properly. He must therefore ensure that the persons responsible for each activity are clearly identified, properly briefed, and provided with the means to receive and hand on their tasks.

Every task is controlled at every stage

To assist in the operation of single point responsibility, a system of 'sign off' control documents (see Chapter 13) is used on particular tasks. As each critical activity is completed a responsible person signs it off and dates it on a control sheet. The sheets are prominently displayed in the control cabins and are continuously monitored and updated. The display allows everyone to see the current state of the tasks involved and spot any task which has either fallen behind or has a 'hole' – a space where an activity has not yet been signed off.

The organization is a blend of the required knowledge and experience

The optimum organization would blend the following:

- *plant personnel*, who possess local knowledge;
- *turnaround personnel*, skilled in planning, co-ordination and work management;
- *technical personnel*, who possess engineering, design and project skills;
- *contractors*, and others, who possess the skills and knowledge to execute work.

In a real situation, the optimum may not be achievable. The turnaround manager, with the assistance of the policy team, must build the best organization possible with the available personnel. To this end, they must be able to assess the existing personnel, recognize its strengths and weakness, and take steps to maximize the strengths and minimize the weaknesses.

Constraints

There is no standard pattern, shape or size for a turnaround organization. For a particular event these characteristics will depend upon the constraints operating at the time. There are many such constraints, all of which have a greater or lesser influence on the form of the organization created. Typical constraints include, but are not limited to:

- the amount of money available to plan and execute the work;
- the size and complexity of the turnaround;
- the timing and duration of the event;
- the availability of personnel;
- the company culture and norms.

The task of the turnaround manager, working within current constraints, is to consider the options available and to design an organization which, in his estimation, best meets the needs of the event. He must then propose this organization to the policy team and inform them of any unavoidable weaknesses and their likely effects.

In the light of any exposed weaknesses, the task of the policy team is to test the proposed organization against their set objectives and to take one of four courses of action, viz.:

- *approve the organization* as proposed, recognizing any weaknesses which may impair its capability to achieve the pre-set objectives;
- *remove the constraints* so that the organization is better suited to achieve the pre-set objectives;
- *change the objectives* to suit the current constraints and organization;
- *change the organization* to better suit the objectives and constraints.

This requires a realistic approach. It is not a rare thing for a policy group to demand the achievement of objectives which are not attainable but to refuse to remove the constraints. The turnaround manager must avoid being pulled into this type of fantasy situation because, in the cold light of a missed objective to which he has committed himself, he will be left to take the consequences, whatever they may be. In the final analysis, the success or otherwise of the organization depends crucially on the balance between constraints and objectives and the creation of a team capable of achieving the objectives.

Scarce manpower resources

Some of the manpower resources required by the turnaround organization may be scarce. While it is true that, in some situations, they may all be 'scarce', what is referred to here are those – e.g. coded welders, versatile instrument technicians, machines fitters – which are scarce by reason of the type of work involved. They may also include specialist vendors such as installers and servicers of digital control systems, or inert entry crews. The work thus involved should be classified as 'critical tasks' and could include, inter alia:

- overhaul of machinery (especially of older models);
- relief stream inspection and registration (often ignored);
- inert entry (especially in vessels with complicated internals);
- high voltage electrical work (which normally requires a licence);
- handling material which may be radioactive, explosive, inflammable or toxic (and which may therefore require a licence);
- coded welding (especially where exotic materials or rarely used techniques are involved);
- calibration of instruments (and electronic and computerized equipment);
- trip and alarm testing (a balance of expertise and local knowledge);
- protective coating and specialist painting (which may look simple but isn't).

Task difficulty, or the need for complex knowledge, specialized trade qualification and legally required licensing, limits the number of individuals who can perform critical tasks or are allowed to. Such personnel have the following characteristics:

- they form the top few per cent of the 'normal distribution' for their skill;
- they are usually highly paid and in great demand;
- they are usually available for a limited period only;
- they are typically focused on a very narrow band of tasks (often on only one).

Because the above personnel are defined as critical, they must be managed with care. Creating a critical manpower resource profile will help. A sample routine for this task would be as follows.

1. For each separate area of the turnaround make out a list of the tasks which will require scarce resources.
2. Using best judgement (remember this is done at an early stage) assign a programme day when the resource will be required (e.g. Day 4).
3. Assign a programme time for the use of the resource (the best that can be done here is probably allocation to the morning, afternoon, evening or night shift).

4. Define the equipment required to perform the task (e.g. TIG weld set).
5. Define any logistic requirements (e.g. a generator, or rod ovens).
6. Define any dependent activities (e.g. radiography, pressure testing).
7. Identify any potential contingency (e.g. a repair or re-weld may be required).

Once the above list has been created the data is organized to create the profile as follows:

1. Create a chart with programme days along the horizontal axis and required manpower resource numbers on the vertical axis.
2. Record the resource requirements for each programme day.
3. Note the requirement for similar work on different tasks and areas for each day.
4. Manipulate the times so that the maximum amount of work is done by the minimum manpower resource.
5. Consider changing the timing of certain jobs (within the confines of the overall schedule) to move manpower from one day to another.

Remember, this is a planning exercise which is initiated in the early days of the preparation phase to give an indication of the likely manpower resource requirements on critical tasks. The first run through will be pretty crude but, as better planning and scheduling information becomes available, the profile can be made more accurate. The manager must exercise judgement on these issues because the scarce resources may have to be procured long before there is any accurate data available – that is their nature!

Once the scarce manpower resources have been identified, the following facts must be established:

- the various providers of such resources;
- numbers likely to be available from each source during the event;
- any factors that will affect resource availabilities;
- the lead time required to hire each resource;
- alternative providers if the primary source dries up;
- the lead times for increasing or replacing resources;
- differentials in pay scales for manpower resources from different suppliers (ignorance of this has caused industrial action during an event because people were not happy about doing the same job as someone else, but for less pay).

Scarce resources may be obtained:

- from within the client company, either on the plant in question or at another location;
- by hiring from another company in the same business as the client (a competitor?);

- from equipment vendors (often overseas and very expensive);
- from a specialist sub-contractor;
- from the main contractor;
- in-house, by providing specific training for the work.

Even if the manpower resources are apparently available, the following measures will minimize uncertainty:

- arrange for the scarce resources at the earliest possible date;
- check all paper qualifications;
- if possible, test competence in the specific task required (welders have to do this all the time!);
- monitor and record performance and retain for future reference.

The lack of one scarce resource on the critical path task can extend the duration of the turnaround, and what may seem to be an issue involving only a couple of people and a relatively small amount of money can have a hugely expensive consequence involving all those people who have to be retained during the over-run (and also some other scarce resources!).

Case study

The organization shown in Figure 7.4 was created to suit the needs of a turnaround. Its form was dictated by management's response to the constraints that operated at the time. The main features were as follows:

A consultant turnaround manager was hired to manage the mechanical duration of the event.

(*Constraint*: the client had no one with the requisite knowledge and experience.)

The plant manager controlled the shut down and start up phases.

(*Constraint*: in line with the company's safety first culture the senior manager decreed that the entire plant must be off line before any mechanical work could begin. The plant manager was the only person with the experience needed to manage these phases.)

A plant-based quality team was created to check and sign off the breaking and re-making of joints.

(*Constraint*: in the past, there had been a significant problem with leaking joints on start up, so this additional team was imposed upon the turnaround manager.)

A large up-grading project, carried out by contractors, was controlled by the company's project department.

(*Constraint*: the senior manager hired the turnaround manager to directly manage only the plant maintenance overhaul and considered the project to be plant improvement.)

The plant was heavily instrumented so the control electrical work was treated as a separate 'area'.

(*Constraint*: the senior manager considered control electrical work to be of a critical skill level that was much higher than that of the mechanical work.)

The company had hired a contractor to handle one of the areas.
(*Constraint*: the company did not have enough internal staff to handle the event.)

The Area 1 co-ordinator also co-ordinated Area 2.

(*Constraint*: shortage of experienced co-ordinators.)

Planning and updating was centralized rather than being done by area planners.

(*Constraint*: insufficient funds to hire the extra staff needed for the area planning teams.)

The main effects on the event of the organization were that the turnaround manager, the plant manager and the project manager did not see eye to eye on a number of issues, most importantly the sequencing of work in the few days leading up to the plant start up. Because none of them was in overall control, their disputes had to be arbitrated by the senior manager, who was not technically trained. He tended to back the plant manager because he was the only 'company man' (the other two having been hired in) and this situation led to a duration over-run.

The quality of work in Areas 1 and 2 suffered because the co-ordinator – who was supposed to deal with both areas – was over-worked and unable to co-ordinate either area effectively. The number of leaks did fall drastically due to the close monitoring of the quality team, and the instrument work – especially the trip and alarm testing during start up – was very successful.

Could the organization have been designed differently to make it more effective? The organization which was created to control the event was heavily influenced not only by the constraints operating at the time but also by senior management assumptions which coloured the decisions about who should do what. As can be seen from this case study, it is not so much the constraints which dictate the final shape of the organization so much as the management's responses to those constraints. Different responses would have created a different organization.

8
Site logistics

Introduction

The business of logistics, in the context of a turnaround, is the organization of the reception on site, the storage or accommodation, the maintenance and mobilization – of every item required for the event. Put simply, the logistic activity is aimed at ensuring that the right thing is in the right place at the right time and in a fit condition to perform its function. Site logistics encompasses all of the non-technical elements of the turnaround (the technical elements being defined as the activities required to carry out the tasks on the work list) even though it deals with technical equipment. It concerns the disposition of the many thousands of items, large and small, that are required by the technical teams to perform their activities in a timely manner. Logistics therefore affects everyone employed on the turnaround.

Because of the above requirements, site logistics must be planned and prepared with a rigour equal to that imposed on technical planning; this is crucial to the success of the turnaround. Poor logistics can ruin the best devised technical plan and thereby prevent the achievement of the turnaround objectives.

The site logistics team

The logistics team is charged with planning and operating the site logistics programme before, during and after the event – and is led by a site logistics officer whose duties include, inter alia, the following:

- drawing up the master site plot plan;
- liaising with plant personnel and the turnaround planning team;
- arranging all outside laydown areas;
- setting up stores and receipt/issue procedures;
- receiving, locating, maintaining and mobilization of:
  - materials proprietary items and consumables,
  - tools and equipment,
  - services and utilities,
  - accommodation and facilities;

- the control of hazardous substances;
- providing for the daily needs of personnel.

The logistics officer reports directly, and refers any operational issues, to the turnaround manager. The logistics team, normally made up of material marshals, storemen and drivers, assists the logistics officer in the execution of the logistics plan and reports directly to him.

The elements of site logistics

There are two elements to be considered. The first is the actual physical object or substance to be dealt with and the second is its current disposition.

Physical objects or substances (see Figure 8.1)

This element refers to all of the items which are required not only to carry out the technical tasks but to take care of the day-to-day needs of the personnel employed. It includes, among other things, the following:

Materials	–	steel, wood, plastic
Proprietary items	–	valves, pumps, electric motors
Consumables	–	welding rods, filters, cartridges
Equipment	–	radiographic, torquing, bolt tensioning
Tools	–	electric and pneumatic hand tools, spanners
Transportation	–	personnel carriers, wagons, low loaders
Cranage	–	tower cranes, mobile cranes, hoists
Utilities	–	electricity, gas, water
Services	–	scaffolding, insulation, water washing
Accommodation	–	offices, stores, conference rooms
Facilities	–	toilets, changing rooms, mess rooms

It is vital that all of these physical items are acquired. It is equally as important that the current disposition of any item be controlled so that, at any given time, the questions listed in the next section can be answered.

The current disposition of the physical objects or substances (See Figure 8.2)

During a turnaround, among the most commonly asked questions are the following:

Disposition stage	*Question to be answered*
Receiving on site	Has it arrived?
Accommodating or storing	Where is it?
Protecting and maintaining	What condition is it in?
Moving around site	Can I have it here, now?
Disposal	Is it off hire?
Recording of current disposition	Where is the proof?
Communicating with others	What is going on?

Typical lists of items and services acquired, accommodated, maintained and supplied by the Site Logistics team include, but are not limited to:

Materials, proprietary items, consumables

- ferrous/non-ferrous metals, wood, plastic, paper
- fabrications, piping, flanges, flat or formed plate
- pumps, valves, motors, instruments and spare parts
- joints, nuts, bolts, studs, washers and pins
- electrical equipment, bulbs, wire and switches
- welding electrodes and wire, solder and flux
- bottled air, oxygen, acetylene, nitrogen and hydrogen
- de-ionized, demineralized and distilled water
- seals, filters and desiccants
- dye penetrant stain and developer
- paper overalls, gloves, overshoes and plastic bags
- lubricants, detergents and cleaning materials
- diesel oil and petroleum

Equipment and tools

- welding rectifiers, generators, compressors
- diesel oil, petrol and water tenders
- welding torches and oxyacetylene torches
- hoists, pulleys, pull lifts and hydraulic jacks
- cables, hoses, ropes, chains and slings
- ramps, barriers, zip-up scaffold and ladders
- bolt tensioning and torquing gear
- on site machining equipment
- inspection instruments and measuring equipment
- radiographic, ultrasound and MPI equipment
- mechanical, electrical and instrument tool kits
- specialist machines, jigs, fixtures and tools
- water washing hoses and lances
- skips, barrows, bogeys, brushes and shovels

Shut down/start up equipment

- special pumps, valves, bobbins tools and fixtures
- safety equipment and protective clothing
- hoses for washing and draining
- barrels and trays to catch liquid spillage
- boxes and bags to catch solid spillage
- detergent for cleaning oil or product spillage
- equipment for cleaning drains and culverts
- skips and trolleys for removing solids
- slip plates, blanking plates, nuts and bolts
- special plates for corrosive product
- temporary platforms, scaffold and ladders
- compressed air and instrument air
- adequate gas, electricity and mains water
- de-ionized, demineralized and distilled water

Cranage and transportation

- large semi-permanent cranes
- small mobile cranes
- hy-ab trucks
- heavy trucks
- flat bed trucks and tenders or trailers
- panel trucks
- small vans
- four wheel drive vehicles
- electric trolleys and fork lifts
- ambulances and fire trucks
- minibuses
- large buses
- motorcycles
- bicycles

Utilities and services

- mains electricity, gas and water
- water and chemical washing, blast cleaning
- inert entries, air moving, lighting, ventilation
- scaffolding, cladding and sheeting
- insulation (lagging and de-lagging)
- in situ machining and cutting
- air-arc, plasma and water jet cutting
- IG, MIG and plasma arc welding
- photogrammetry, laser aligning, acoustic ranging
- ultrasound measurement and thermal imaging
- introscopy and remote cameras
- radiography, ultrasound and MPI inspection
- bolt tensioning, torquing and testing
- ceramic metals and metal spraying
- personnel lifts, hoists and manveyors
- cleaning and scrap removal

Accommodation and facilities

- temporary offices, messing rooms, stores, locker rooms, washing and toilet facilities
- induction, first aid and safety cabins
- temporary pathways and duck boards
- electric supplies, wiring and connections
- water supplies, pipes and connections
- telephones (inc. lines), radios (including batteries)
- computers, photocopiers and facsimile machines
- office furniture, fittings and equipment
- messing facilities and provision of meals
- paper towels, soap and toilet rolls
- fire and toxic alarm facilities
- transportation of personnel to, and from work
- food and beverages
- cleaning services for accommodation

Figure 8.1 Physical objects and substances

The following issues must be considered

Receiving on site
- visually check quantities against purchase order
- query partial deliveries with procurer
- visually check for damage or deterioration
- arrange for technical inspection
- quarantine nonconforming items and arrange for concession or return to sender
- inform procurer of disposition
- cross check order, delivery note and invoice
- pass documents to procurer for processing
- record reception in ledger

Protect and maintain
- provide fire and toxic refuges
- provide first aid posts and ambulances
- provide secure storage for valuable items
- protect items against corrosion and impact
- provide clean conditions for electronic gear
- check shelf life of perishable items
- service and lubricate mechanisms where required
- use correct storage procedures for gaskets, seals and other delicate items
- use good handling procedures to avoid damage
- repair or replace any damaged item
- provide protection for items stored in the open

Dispose of hired equipment and services
As soon as possible because, during every extra hour on site, they:
- will cost money
- will take up space
- will be liable to be damaged
- will be liable to be misused
- could represent a hazard

Make arrangements with suppliers so that:
- your liability ceases the moment the equipment is off-hire
- supplier is responsible for removing it from site

Communicate current disposition
The logistics officer must communicate effectively with other key personnel to inform them of:
- the arrival of items and services
- current condition (quarantined/damaged?)
- current location (stores/lay down area)
- current disposition (stored or in use)
- estimated time of arrival at on-site location
- unpredicted shortages (and strategy to overcome)
- damage or deterioration (and steps being taken)
- minimum stock levels (reorder of critical items)
- items and services off-hire (final cost)

Accommodate or store
Provide the following:
- adequate accommodation for people
- adequate storage space for small items
- lay down areas for large items
- locations for services such as water washing, scaffolding, insulation and other service equipment
- secure areas for hazardous substances
- locations for cranes and other heavy equipment
- locations for rectifiers, generators, compressors
- safe parking space for all vehicles
- rubbish skips, bins and bags

Move items around site
- control the movement of all vehicles
- nominate approved routes for the movement of personnel, vehicles and equipment around the site
- provide safe routes for hazardous loads
- prohibit the use of hazardous routes
- prohibit permanent blocking of approved routes
- if routes are blocked, provide alternative routes
- procure a priority movement list at busy times
- avoid causing damage to items being moved
- lay on vehicles for overtime and shift work
- protect, maintain and regularly refuel vehicles
- protect pipes and cables which lie across roads

Record current disposition
The logistics officer must know where every item is at any given time. There are thousands of items spread around many locations in various stages of use. Therefore, the site logistics officer must have an effective procedure which shows:
- date received and present location
- items damaged, quarantined or returned to sender
- end user informed of arrival
- date issued, quantity and receiver
- remaining stocks and minimum stock levels
- hired items and services off-hire date

Any other issues

Figure 8.2 Current disposition

The aim of the logistics officer and his team is to prevent the need for such questions (by knowing the numbers and types of physical items that are on site and the current disposition of each), and where that is not possible, to be able to answer them and other related questions with the minimum of delay.

In order to organize such a vast amount of information, the logistics officer requires some sort of conceptual model into which most, if not all, of the data required to effectively supply the necessary items for the turnaround can be inserted. Supported by various schedules and procedures, the model used is the plot plan.

The plot plan

The plot plan is one of the most important documents of the turnaround. It is, to the non-technical aspect of the event what the schedule is to the technical aspect. A good plot plan will afford the logistics officer the opportunity to control supply. A bad or non-existent logistics plan may lead to confusion at best and chaos at worst. This, in a business context, leads to over-run, overspend and a negative effect on quality. When the plan is completed it will display the location of every important element on the turnaround.

Purpose

The purpose of the plot plan is to ensure the safety, availability and effective mobilization of every item on site. To achieve this it is necessary to organize the current disposition of every physical element involved (as already discussed in this chapter).

Preparing the plot plan

To prepare the plot plan, the logistics officer must survey the site and gather a large amount of information from plant and other personnel. There may exist a plot plan from a previous turnaround and, if so, he may use it as a basis for his current plan. Failing this, there should be a basic plot plan of the plant, showing the position of all the critical elements. Some such plans are very detailed while others contain only a minimum of information.

Where a plot plan does exist, the logistics officer should not assume that it is accurate. Many plans are the original ones issued when the plant was commissioned (perhaps as much as twenty-five years or more in the past) and which have not been updated to reflect the many subsequent modifications. In the worst case there will be no plot plan and the logistics officer must create one.

It is therefore in the best interest of the logistics officer to survey the plant and gather information – to ensure that the basic turnaround plot plan will be accurate. Finally, it must be drawn to a scale large enough for individual elements (located via a grid-reference system) to be identifiable.

The stages

There are three stages involved in producing a plot plan, viz.:

1.   drawing up the basic site map;
2.   drawing in the turnaround overlay;
3.   issuing the plot plan.

Drawing the basic site map

Whether the basic site map is provided by the plant or created by the logistics officer, it must show such important features as:

The perimeter of the plant and the boundaries of the available land (if any) surrounding it – because a great deal of work will be carried out in a short space of time, the logistics officer must have a clear indication of what land is available to lay down the goods and services required.

All major items of plant equipment and connecting pipe work – in the correct configuration to ensure that available land adjacent to hazardous areas, areas of special function, or areas with difficult or restricted access, are identified.

All roads (in particular, public ones) which run through the site and/or the surrounding land – so that any interface with the public can be made safe.

All access to site and site roads – such as personnel, goods entry and emergency evacuation gates. During the event the number of people and the volume of goods and services entering and leaving site will be many times greater than normal and extra gates may have to be created.

The location of all fire assembly points, toxic refuges, fire fighting equipment, alarms, site telephones, emergency showers and eye baths – to ensure that such provision is adequate and that all turnaround personnel can be shown the locations of this equipment as part of the site safety briefing.

Any other permanent or unusual feature – which could affect the site logistics (e.g. heights of pipe bridges).

Once this basic map has been completed the logistics officer then has the task of applying all of the data which will indicate the disposition of the site, during the turnaround, to the basic map. This is called the *turnaround overlay*.

The turnaround overlay

The logistics officer accurately marks the position, and codes the function (by colour, numbering, or other means), of every element added to the basic map for the purposes of the turnaround. Typically these will include, but not be limited to:

●   any non-load-bearing surfaces, where the siting of cranes etc. is prohibited;
●   any areas or roads where access is prohibited;

- approved vehicle routes, with the direction of traffic flow (especially one way systems);
- areas designated for turnaround stores and quarantine compounds;
- areas for hazardous substances – catalyst, chemicals, paint, oil, fuel;
- foul laydown areas for contaminated items;
- water washing and chemical cleaning bays (water supply, drainage and access);
- clean laydown areas for new and overhauled items;
- laydown areas for large fabrications;
- marshalling areas for scaffolding and insulation contractors;
- parking areas for welding rectifiers, air compressors, electrical generators and fuel tenders;
- areas for accommodation, mess rooms, changing rooms and toilet facilities;
- temporary cabling and piping routes for utilities;
- parking areas for cranes and heavy equipment;
- parking areas for site vehicles;
- parking areas for private cars and buses;
- location of additional temporary safety equipment;
- sites for the turnaround control, induction and safety cabins.

A special additional category is required if any ground is to be excavated. In this case drawings of the site underground services must be obtained and verified – and the relevant information overlaid on the site map.

In addition to the above, the logistics officer must mark in any other areas or special features local to the particular plant being worked on. Once this is done the plot plan is complete.

Issuing the plot plan

Throughout the creation of the plot plan the logistics officer must consult with the turnaround manager regarding its feasibility. When it is completed, he must submit it formally – for discussion, approval and action – to the turnaround manager who will then discuss it with the plant manager and his staff, with the safety officers for the site and for the turnaround, and with any other interested party. Any resulting amendments are then carried out and the final plan published, copies being issued by the logistics officer to the following personnel (at least):

- turnaround manager and area engineers;
- plant manager and area superintendents;
- turnaround safety officer;
- turnaround planning officer;
- site security;
- contractor managers;
- any other nominated key persons.

It is the responsibility of the key personnel to study the plan, become familiar with its features, comment upon any circumstance or feature which may affect its usefulness and then brief their staff to ensure that everyone has a good working knowledge of it. The plans should be displayed prominently at key locations in the turnaround control offices. Briefing and display are very important, because once the event starts everyone on site must conform to the requirements, restrictions and prohibitions of the plot plan. Failure to do so may cause accidents, delay, over-run or overspend. Segments of the plan may be photo-copied to aid other specific planning or execution activities but the copies must not be amended.

Site logistics is a large and important element in the overall turnaround plan. As with all the other elements it requires accurate planning and validation.

9
The cost profile

Introduction

Creating a budget for a project is relatively straightforward if the exact work scope is known (and will not change over the life of the project) and the unit costs of resources, goods and services are also known. Unfortunately, this is almost never the case with turnarounds. The budget will be hedged around with uncertainties (some of which have been enumerated in previous chapters), the chief of which being uncertainty about the actual work scope (see Chapter 3) and allowance for contingencies. Cost estimating should therefore be performed (for major events, at least) by an experienced cost engineer.

Why estimate a cost profile?

The purpose of putting together a cost estimate or a budget is to ensure that the estimated costs of the event are known as far in advance as possible. This will allow those responsible (mainly the policy team) as much time as possible to optimize the constraints on the budget – by pricing each major element of the turnaround and holding costs to a minimum. Additionally, the costs must be organized in such a manner as to allow expenditure to be monitored and controlled (see Figure 9.1). This demands accurate costing and this in turn depends upon exercising a number of disciplines, namely:

Achieving work definition of the highest possible quality – given the circumstances surrounding the event. Realistically, not all work can be definitively specified. There are unknowns because the items to be worked upon are mainly concealed from view when the plant is running (which is usually the time when planning is carried out). The planner is sometimes reduced to guesswork based on previous experience (which is not always sufficient). Where uncertainties exist or assumptions have been made, they should be specified.

Closing the work list on a specified date – usually four to six months before the start, for a major event. The line must be drawn somewhere to allow the estimate to be made. If the influx of work requests is is never-ending a budget cannot be fixed.

Separate costing of all work requested after the work list closure date. Realistically, work requests *will* continue to arrive after the closure date, due to the fallibility of people and equipment. A separate 'Late work order system' should handle this. The work is planned out only after the work on the approved list has been planned (unless there are special reasons why it should be dealt with earlier and these can only be generated by agreement between the turnaround and plant managers), and it should be costed separately on a late order budget sheet.

Allocation of a unique cost code to the turnaround. If this is not done it will be impossible to track down the event's costs.

Turnaround manager approval of every item of expenditure – any number of people may otherwise allocate cost to the turnaround and control over expenditure would be lost.

The inclusions

Having imposed these disciplines, it is then necessary to define what elements shall be included in the cost estimate. These should include, but not be limited to the following.

Turnaround planning and management – this may be charged in a number of different ways, as a fixed and firm lump sum, as a management fee plus reimbursable hours for staff such as planners, or as totally reimbursable for the complete turnaround team.

Local labour – covering the plant personnel who will be engaged on the event. Some companies do not include these costs in their turnaround budgeting, but ignoring them leads to an over-optimistic estimate of the total cost.

Contractors – the costs should include those of all the contractors employed. On a major turnaround there could be thirty or more, from the large multi-skill companies who will carry out substantial packages of work, through the medium sized ones who provide services such as scaffolding and water washing, down to the very small one-task specialists. Care must be taken that *all* contractor costs are recognized at an early date, especially the hidden elements. (In one actual case a contractor presented the client with a bill for 'disbursement' costs of $135,000 which, although not stipulated clearly during negotiations, were held by the contractor to be implied in the contract. The contractor threatened litigation and the client settled out of court.)

Stock and miscellaneous material – stock costs being those of recognized spares, miscellaneous costs those of all materials bought to repair defects etc.

Equipment purchase and hire – equipment can only be properly identified if work methods are specified in detail. Communications equipment – phones, faxes, photo-copiers, computers, printers etc. – is often overlooked.

Accommodation – for a major turnaround on a plant with limited permanent accommodation, the cost of offices, accommodation cabins, mess rooms, changing rooms and toilets (together with their cabled and plumbed-in services and drainage) can be substantial (even toilet rolls cost money!), especially if the cabins have to be sealed units that may act as refuges during a toxic release. The cost of the fitments (tables, chairs etc.) must also be taken into consideration.

Utilities – a substantial amount of electricity, water, various gases, oil, petrol and diesel will be consumed. (It all has to be paid for.)

Contingencies – as in planning, so in budgeting. Contingency costs are a grey area in which educated guesses based on previous experience (if there is any) are the most accurate estimates available.

The inclusion of the above elements and (with the exception of contingency costs) the accurate calculation of the cost of each will generate the most accurate budget possible.

The exclusions

There are other costs which are normally excluded from the budget. Whether they are, in any particular case, will depend upon company culture and the prevailing circumstances. They are listed below.

Any project work which will be handled independently – although it may be carried out at the same time as the turnaround and needs to be integrated into the schedule, the cost is not calculated as part of the turnaround estimate.

Pre-ordered materials and proprietary items – these costs will properly form part of the annual maintenance budget and should not be dumped into the turnaround budget.

Work requested after the closure of the work list – as explained previously, the work list must be frozen before an accurate cost estimate can be made. Any work requested after this, if accepted, must form part of the late work order system, and be costed separately.

Emergent work – whether it is generated as extra work, additional work, or by a change of intent, this should be covered by the contingency element of the budget.

Creating a cost estimate

The cost estimate is built up in three basic stages – the ball park estimate, the proposed estimate, and the final estimate. These are developed over a period of time, the accuracy of the estimate being progressively refined as more information is made available.

Stage 1, the ball park estimate

The first stage is to create a ball park estimate using the initial work list. This is typically done six to nine months before the event and has an accuracy of no greater than plus or minus 20 per cent. It is an early warning system that is used to give the policy team an indication of the likely magnitude of the final cost estimate. For example, if the budget allotted to the turnaround is $2,200,000 and the initial cost estimate is $2,800,000 then it is clear that even the lower band (–20 per cent) of the estimate, at $2,240,000, exceeds the budget – they are not in the same ball park. The higher band at $3,360,000 is radically more expensive. On the other hand, the initial estimate may indicate that the predicted expenditure is well within the allotted budget. Either way, the policy team knows at the earliest possible time what the final outcome is likely to be and, if this is unsatisfactory, has the time to take remedial action.

The initial budget is built up by using one or more of the following techniques:

Analysis of existing data – such as budget and cost reports from previous turnarounds on the plant in question or on similar plants. The questions to be asked here are:

- Was it a similar event with a similar work scope?
- Can the budget be re-used as it is, or with minor modifications?
- If not in total, can any elements of the budget be re-used?

Analysing the major tasks on the work list – and determining, from plant records, whether similar tasks have been done in the past. If so, how much did they cost?

Aggregating bulk work – and applying existing labour rates to them. For example, if there are three hundred valves of various sizes and degrees of complexity, calculate an average time per valve in hours and multiply it by an average hourly rate.

Using any existing quotes – some tasks may have already been costed.

Using available norms – some companies have calculated norms – based on an average of many similar tasks done in the past – for different types of work.

Using weekly turnaround control team costs – these will typically have been calculated and agreed during the initiation phase of the turnaround.

Applying hourly rates for specialist skill groups – these will readily be provided by vendors.

The actual strategy adopted for putting the initial cost estimate together will depend on how much hard information is available. What if there is none, however, apart from the initial work list? The response to that particular challenge is to use a technique which is based on a reasonable ability to estimate averages.

The quick and dirty estimate

In situations where there is little or no historical data, the following technique can be used. It is based on four principal (and not unrealistic) assumptions, namely that:

1. the initial work list is known,
2. basic man hours can be calculated,
3. current labour rates are known or can be accurately estimated,
4. on turnarounds, manpower costs typically represent around 30 per cent of total costs. (If better local data, regarding this ratio, exists then it should be used.)

The steps in the technique are as follows:

1. *Select the most experienced team* – from those who have experience on either turnarounds or maintenance (this would include the turnaround planners).
2. *Estimate the man hours for each of the major tasks* – purely from experience, or from past practice, or by simple calculation (analysing the task into steps and estimating, per step, the number of men times the number of hours required).
3. *Estimate the man hours for each of the small tasks* – a simpler version of Step 2, usually made easier because small tasks can usually be grouped so that an estimate for one of the tasks in any group can serve for each one of the tasks in that group.
4. *Estimate the man hours for bulk work* – an even simpler task. A large number of similar small tasks (on valves, bursting discs etc.) can be aggregated and the average time per task estimated. In any particular case, the actual number of tasks is then multiplied by that average time.
5. *Add together all of the man hours* – from the above three categories to give basic total man hours.
6. *Multiply the basic man hour total by a productivity factor* – to get the estimated total man hours. This factor takes account of non-productive hours and is normally between 1.4 and 2.0, depending upon how effective you judge the local labour and supervision to be.
7. *Estimate (or, if possible, calculate) an average hourly cost for labour* – this will be based on the ratios of specialist, mechanical, instrument, electrical, civil, semi-skilled and unskilled labour used.
8. *Multiply the estimated total man hours by the average hourly cost* – to give the basic manpower cost.
9. *Multiply the basic manpower cost by a factor of 3.3 (or whatever factor is most suitable to local conditions)* – to give the basic overall cost of the turnaround.
10. *Consider the worst case scenario* – estimate the overall cost if all the circumstances and conditions assumed turned out, in practice, to be as adverse as they could be.
11. *Consider the best case scenario* – estimate the overall cost if everything went perfectly.
12. Multiply the overall cost derived at Step 9 by four and add to it the figures estimated at Steps 10 and 11, and then divide the total by six. The new total arrived at is the ball park cost estimate for the turnaround. Obviously this is a very rough figure, but (i) there may be no other way of arriving at a ball park estimate, and (ii) the accuracy sought is only around plus or minus 20 per cent.

This figure, and its estimated limits of accuracy, should be presented to the policy team as early as possible.

The proposed cost estimate

As the preparation phase proceeds and more hard information is obtained on costs, the estimate is refined – using actual and calculated costs – to the point (typically two months before the event) where it can be presented to the policy team as a proposal with an accuracy of about plus or minus 5 per cent. In preparing an accurate cost estimate it must be ensured that:

- every known relevant item of cost is included;
- unit prices are up-to-date and accurate;
- either contingency allowances, for an estimated amount of emergent work, are included or, if not, the exclusions are clearly stated;
- all known or estimated non-productive time is factored in;
- all assumptions are stated.

One of the best investments the turnaround manager can make is to engage (as suggested earlier) the services of a cost engineer or quantity surveyor to put together the cost estimate (especially the proposed cost estimate). This individual can:

- produce an accurate cost report;
- examine contracts for 'hidden costs';
- challenge work practices to reduce costs;
- present the cost estimate in detail to the policy team;
- create measuring tools for monitoring expenditure;
- measure performance and expenditure before and during the event;
- produce regular live reports and forecasts to allow the turnaround manager to track costs on a daily basis.

The approved turnaround budget

When the proposed cost estimate, based on the approved turnaround work scope, is finalized it is presented to the policy team for analysis, discussion, decision and action (see also Chapter 6, Optimizing the turnaround plan). If the estimate is less than the allotted budget figure, the saving is quantified and formally recorded. If the cost is greater, the excess is quantified and the following options are explored in order to bring the costs back within the budget figure:

- eliminate work from the work list or reduce the complexity of certain tasks;
- defer some tasks to a later date or another planned outage;
- challenge the unit costs of labour etc.;
- reconfigure turnaround elements.

The alternative to the above is that the policy team could increase the budget. When the estimate is finalized and approved by the policy team, it is adopted as the approved budget for the turnaround.

During the event, the key members of the turnaround team are given specific responsibilities for controlling costs. Figure 9.1 shows a checklist of actions necessary to ensure that the reponsibilities are met.

Check list of actions required to control costs

Turnaround manager
- brief key personnel on the budget and cost objectives
- set area engineers a cost saving initiative
- agree a cost review timetable with the QS
- vet all variations to contract and approve if justified
- vet variations to labour levels and approve if justified
- vet overtime requests and approve if justified
- vet changes of intent and approve if justified
- vet contractor claims and approve if justified
- review extra and additional costs with area engineers and ensure they are separately recorded
- audit area engineers' cost performance
- review cost report and forecast with QS and engineers and take action to correct cost over-run trends
- report cost issues to the policy team

Area engineers
- study the area plan and cost profile to gain a deep understanding of the distribution of costs
- brief the area team on the budget and cost objectives
- control manpower levels and reduce if possible
- vet hire extensions and approve if justified (on major items, check with the manager)
- investigate overtime claims and refer to manager
- investigate daywork claims and authorize same day
- investigate contract variations and record them
- investigate extra work costs and refer to manager
- investigate additional work requests and their cost implications and refer to manager
- review cost report and forecasts with manager and QS
- proactively search out cost saving opportunities

Engineers running contractors
- issue ITB's early to give contractors enough time to develop a realistic price
- ensure ITB's contain all data necessary to allow contractors to identify all work required
- ensure the site is ready for the contractor
- ensure the contractor arrives and leaves to plan
- strictly control all variations to plan and cost them
- minimize contractor waiting time
- report conflict with other work to the turnaround manager as soon as it occurs
- report any contract overrun to the turnaround manager at the earliest possible date
- ensure contracts are properly closed out and final price known at earliest possible date

Quantity surveyor/cost engineer
- produce a cost profile and publish approved budget
- produce regular cost report and forecast, discuss with manager and give early warning of negative trends
- produce financial instructions for contractors
- check contractors daywork claims against contract
- check work measures on scheduled rate contracts
- assist manager to settle contractor claims
- advise area engineers on changes to reduce costs
- check hire dates of all major hired equipment
- collect and collate data from all cost centres
- audit site work to ensure best practice and no waste
- report any cost generating issues to the manager
- write a final cost report with recommendations for future cost saving initiatives

Co-ordinators and supervisors
- do not carry out any extra or additional work on verbal request – ensure it is written down
- do not verbally request extra or additional work from contractors – write it down and report it to the area engineer on a daily basis
- log the use of additional material and hire equipment
- log 'off-hire' dates for equipment and ensure they are met or bettered (under normal circumstances)
- request overtime only when it is unavoidable
- keep tight control of material issued in their area to eliminate waste – which attracts costs
- keep tight control of equipment in their area to avoid damage – which attracts costs
- provide any information requested by the QS

Turnaround workshop manager
- ensure he has an approved work list
- provide unit prices for standard tasks
- ensure any extra/additional work is logged and approved by the turnaround manager
- if overtime is required, it must be approved by the turnaround manager
- if any task is taking longer than planned, inform the turnaround manager
- provide a daily record of costs for the turnaround manager

Figure 9.1 Cost control

Case study

A chemical company ran three turnarounds on adjacent plants at the same time. It appointed an event manager for each plant and a turnaround manager to take overall control of the total project. The latter was given special responsibility for creating an integrated cost estimate, one which would include and combine the costs for all three plants. The company operated a computerized maintenance management system (CMMS) and there was only a limited number of 'experts' in the company who could use it with any degree of competence. There were no fewer than forty two people who had the authority to input work and material requests into the system during the normal operation of the plants. Most of the material for the turnaround had been pre-ordered under a number of different cost codes and much of it was mixed in batches with materials for routine maintenance.

The event managers did not concern themselves with trying to control costs because they saw this as the turnaround manager's job. The latter could not stabilize the cost estimate because the costs were lodged in so many different places and obscured by being mixed in with other costs. The company seconded a number of the CMMS experts to help the turnaround manager to interrogate the system.

The cost estimate was never stabilized. It would rise sharply as a result of one of the CMMS experts suddenly discovering large costs which had been hidden within the system. Then it would fall when other costs were found to be incorrectly allocated to the turnaround. This situation continued throughout the preparation phase so that the final cost estimate for the events was fictitious. The situation also continued throughout the execution phase and was further complicated by the additional cost of a large amount of emergent work and the decision to take some planned work out of the events in order to try to 'balance the books'. Six months after the turnaround was completed the final bill had not been calculated because claims for payments were still coming in and there were several disputes between the plants as to who should bear which costs.

What were the basic errors committed in this case?

The senior management did not have a policy for cost control. They assumed that everyone would know what to do and would co-operate in controlling costs. The turnaround manager did not operate a single cost code system for the event and as there was no penalty for ordering materials without approval, many people did just that. The event managers acted irresponsibly by not controlling their own costs. Ultimate responsibility lay with the policy team who should have addressed the cost issue at their first meeting.

10
The safety plan

Introduction

A turnaround is a hazardous event. It introduces a large number of people into a confined area, to work under pressures of time with hazardous equipment. In recognition of the greater risk of loss, the targets set for safety on a turnaround must be uncompromising – zero accidents, incidents, fires etc. In order to meet these requirements the system of working must be equally uncompromising in its approach to safety.

It is the responsibility of the safety officer, leading a team made up of key personnel, to produce a safety plan which will ensure that all relevant matters are addressed. It would be beyond the scope of this book to enter into a full discussion of the nature of such a plan (the documentation for which can run to over a hundred pages for a major event). The aim here is to define the critical items that come within the purview of the turnaround manager. To begin with, however, it would be useful to define the term 'hazard'.

What is a hazard?

A hazard is any condition, act or event which exposes people, property or the environment to some form of loss, i.e. to:

health – exposure to any substance, noise or other element which causes temporary or permanent illness or impairment of any bodily function;

life and limb – any accidental or planned occurrence which causes death or injury to the human body;

property – any accidental or planned occurrence which causes destruction of, or damage to, plant, structures, buildings, equipment or personal property;

environment – any accidental or planned emission which pollutes the atmosphere, land or waterway in the vicinity of the plant and beyond.

Factors which contribute to exposure to hazards and consequent loss can include:

lack of proper planning and preparation – not thinking the job through;
lack of awareness – ignorance of the hazards;

| *lack of care and attention* | knowing the hazards but ignoring them; |
| *incorrect motivation* | – putting productivity before safety. |

Thus it can be seen that the management of safety is not an academic exercise. It is about real people. The effectiveness of the safe system of work will, in the extreme case, determine whether the people who entrust their safety to that system live or die.

In order to minimize the risk of loss, the turnaround manager has a number of safety strategies that he can call upon. Also, there are basic principles adherence to which will serve to protect people, property and the environment. The most fundamental of these is the safety chain.

The safety chain

There is a chain of responsibility which runs vertically through the turnaround organization and its sole purpose is to ensure that those who form its links understand what their responsibilities are and the resulting actions that have to be carried out. The links in the chain are:

Manager > Engineer > Supervisor > Worker.

Manager

The turnaround manager must provide a safe working routine and ensure that everyone who is employed on the event is properly briefed on that routine before they work on site.

Engineer

The engineers who are responsible for managing areas of the turnaround must analyse those areas, and the tasks that are to be performed in them, in order to expose, and either eliminate or guard against, any hazards associated with the tasks. The engineers must also ensure that the teams working on the area are briefed daily on safety.

Supervisor

The supervisors must ensure that the workplace is safe at all times, tools and equipment being used are fit for the purpose, and those who carry out the tasks are competent to do so. The supervisors must control the permits in their areas and carry out daily safety briefing.

Worker

The workers must follow all safety instructions and not commit any act, or work in any set of conditions, that would endanger their own or anyone else's life. They must report all unsafe acts and conditions to their supervisor or the safety team so that they may be eliminated.

There are many more safety principles and systems relevant to a turnaround, but the safety chain is the backbone of safety. Note that the safety officer is not included. This is because safety is the responsibility of line personnel –

not the safety department. The safety chain forms an important part of the overall safety communications network.

The safety communications network (see Figure 10.1)

Because of the complexity of the event, it is essential that clear lines of communication on safety requirements are established. Figure 10.1 gives an indication of how many people are involved and just how complex the communications network – the overriding purpose of which is to ensure the safety of the event's workforce – can be. The central spine of the diagram shows the safety hierarchy responsible for setting the safety policy and ensuring that everyone adheres to it. The safety chain is defined,

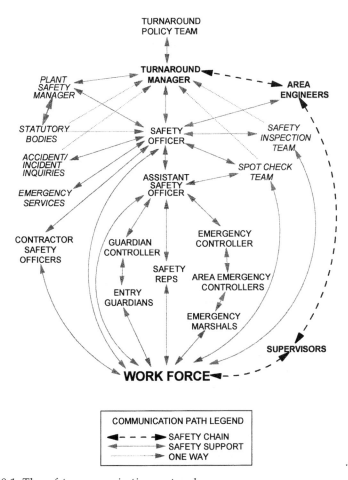

Figure 10.1 The safety communications network

us are the individuals, teams and organizations who can influence the turnaround. The other functions on the diagram, involving the safety team, will be dealt with later.

The safe working routine (see Figure 10.2)

When the turnaround manager, or any person nominated by him, sends workers onto a plant to carry out a task, he must ensure *beforehand* that the hazards associated with the task have been taken into consideration. He must then ensure that steps are taken to eliminate the hazards or to protect the workers against them. There are two categories of hazard on a turnaround, viz:

- a pre-existing hazard, so called because it is inherent in the plant, e.g. hot pipes or toxic substances;
- an induced hazard, so called because it is introduced by the very act of performing the task – e.g. by the use of burning torches.

The safe working routine assesses the following elements:

- the permit to work;
- the workplace environment;
- the worker;
- the task specification;
- materials and substances;
- tools and equipment.

Figure 10.2 offers a number of checklists – which are indicative, not exhaustive – for applying the safe working routine. On one of the above elements, the task specification, a specific hazard assessment must be carried out.

The hot spot inspection

The hot spot inspection deals with pre-existing hazards and is concerned with making the site safe for the turnaround workers. Well before the event starts, the safety officer and the turnaround manager should carry out an inspection to expose any existing hazards (some of which will not be considered hazards during normal operation but will become so because of the nature of the event). The inspection seeks to expose any condition which could pose a hazard during the event. Hot spots could be:

- non-load-bearing surfaces;
- open ditches and drains (especially beside roads);
- plant which will remain live during the event;
- corroded or damaged structures (especially platforms);
- oily or greasy walking surfaces;

Check list of required activities

Permit to work

To ensure safe working a permit must:

- be issued for all tasks and cross referenced to WO
- be specific and relevant to the task being done
- be issued for a specific period of time
- specify, in detail, the work to be done
- specify known hazards and preventive measures
- specify checks to be done and who is to do them
- state if isolation procedure is adequate
- be issued and received by competent people
- be properly handed over at shift change
- be properly handed back on completion of task

Supervisors must spot-check to ensure permit conditions are being adhered to by workers

The worker

The supervisor must ensure that the people who are going to carry out the work:

- are trained and experienced at doing it
- are briefed on safety requirements
- are briefed on the requirements of the task
- understand the permit to work
- understand the task that is to be done
- have the correct materials, tools and equipment
- have been asked for any safety concerns
- feel confident about doing the work
- are wearing the correct protective clothing
- are supplied with safety equipment specified in the permit to work

Materials and substances

- always ensure that any blanking plate is the correct material for the process fluid involved
- ensure that only specified materials are used
- do not specify or use hazardous substances unless it is unavoidable
- if hazardous substances are used, personnel involved must be properly protected
- before using any new substance, analyse it for any potential hazards
- keep a current register identifying substances hazardous to health (COSHH regulations)

Safe workplace environment

A supervisor must check the workplace to ensure:

- access and egress safe and adequate for the task
- area is free of dirt, debris and spillage
- area is not contaminated by any substance
- atmosphere is free of dust or noxious substance
- temperature and noise within approved safe levels
- work above/below/beside the work area is safe
- enclosed spaces are adequately ventilated, illuminated and there is an escape route
- where an unacceptable environmental condition cannot be eliminated it is guarded against
- where an unacceptable environmental condition cannot be guarded against – the job is stopped

The task specification

The task specification must:

- specify a safe method for doing the job
- not assume local knowledge
- be easily understood by those who use it
- specify correct material, equipment and services
- identify known hazards associated with the task
- specify safety precautions for known hazards
- include reference to good housekeeping and care for the environment
- identify the use of hazardous substances
- specify decontamination of foul items

Tools and equipment

- ensure all tools and equipment are undamaged, serviceable and adequate for the task
- discard any tool that is damaged during the task
- ensure air tools have the correct hoses and fittings and that they are periodically inspected
- the use of air equipment to pressure test is forbidden unless written permission is obtained from a competent senior manager
- be aware of the hazards involved in large heavy crane lifts and pre-plan the lift
- ensure that lifting equipment (including beams) has current test certificates

Figure 10.2 Safe working routine

- tripping hazards;
- product drips or emissions (especially acid or toxic substances);
- airborne particles or substances (potential eye injuries);
- restricted access or egress;
- protrusions on walkways (especially at a height);
- ineffective illumination of enclosed structures;
- damaged insulation (especially if it is asbestos based);
- narrow roads (could make transportation or siting of cranes hazardous);
- any other observed unsafe condition.

NB The above list is not intended to be exhaustive.

As a result of the hot spot inspection a list is drawn up of tasks (erection of barriers, cleaning of surfaces etc.) needed to eliminate as many of the inherent hazards as possible. Any hazard which cannot be eliminated must be marked up (in red) on the plot plan as a hot spot. The list of hot spots should be transmitted to the plant team and, in particular, to the permit to work issuers who should be requested to write in special instructions to guard against the hot spot hazards.

Task hazard assessment (see Figures 10.3 and 10.4)

During the turnaround, the worker will interact with all other elements defined in the safe working routine through the medium of 'the task'. It is therefore incumbent upon the responsible person (in this case the preparations engineer) to assess the task in a formal manner. The assessment is, in reality, an investigation of an accident before it occurs, a 'pre-mortem' if you like. It is used to predict potential hazards on a job, to define the types of loss that could result and to specify precautions to be taken to eliminate or guard against those hazards.

Task hazard assessment is a team activity because one person, working on his own, will consider the job only from his own perspective. The types of task assessed could be major ones, multi-level, complex, unfamiliar, or those with a history of accidents. It is also sensible to carry out a generic assessment of bulk work. The assessment process has four steps:

1.   Specify the main steps of the task.
2.   Identify the hazards associated with each step.
3.   Define the potential loss associated with each hazard.
4.   Detail the precautions necessary to protect against the hazard.

Figure 10.3 shows each of the four steps in detail. Figure 10.4 is an example of a completed task hazard analysis (THA) proforma.

Detailed steps in the task hazard assessment

1. Define the main steps

- select a team with a mixture of engineering, safety, and local plant knowledge and experience
- list the tasks to be assessed from the following:
  - major tasks
  - multi level tasks
  - complex or unfamiliar tasks
  - tasks with a history of accidents
- brief the team on each task
- select the first task to be assessed
- list the basic steps of the task
- check and recheck the steps with the team
- write the steps down on the THA proforma

Note: When developing the basic steps use enough detail to clarify the step but not enough to obscure it. This is best done by following the natural steps of the job

2. Define associated hazards

Does the basic step involve:

- multi level working?
- multi task working?
- working at a height?
- working in an enclosed space?
- loosening, unbolting, moving or lifting?
- welding, burning, grinding, blast cleaning?
- work with damaged or corroded items?
- demolition, generating debris or dust?
- removal of scaffold, walkways or supports?
- noisy, cold, hot, wet, dirty or tiring conditions?
- proximity to live plant?
- complex or unfamiliar activities or equipment?
- any natural environmental hazards?
- interface with public roads, traffic or facilities?
- interface with other plants or companies?

Define the associated hazards for each basic step and record them on the THA proforma

3. Define the potential loss

For each hazard, define the potential loss

Can any person:
- slip, trip, overbalance, fall over/through anything?
- get trapped in, between or under anything?
- be struck or contacted by anything?
- have anything fall on them?
- strike against/come into contact with anything?
- absorb (through skin) inhale or swallow toxins?
- be exposed to radiation, gas or corrosive fluids?
- be burned, electrocuted, frozen, drenched or covered in dust?
- cause a fire or explosion or other destruction?

Is plant or other property in danger of being burned, damaged or wrecked by explosion? Will any emission caused pollute air, land, stream, river, sea or ocean? Record potential loss for each hazard on the THA proforma

4. Specify precautions

Prevent the potential loss, i.e.:
- eliminate the hazard
- eliminate the job step
- use an improved method

If the potential loss cannot be prevented, then guard against it i.e.:
- erect extra supports or scaffold
- erect protective barriers
- post warning signs
- use special protective clothing/equipment
- employ specialist contractors
- limit exposure time
- reduce frequency of tasks
- give personnel extra briefing

If the hazard cannot be eliminated or loss prevented, do not do the job!!

Record the precautions on the THA proforma

Figure 10.3 Task hazard assessment

The safety team (see Figure 10.5, and also refer to Figure 10.1)

The size and composition of the safety team will reflect the company's commitment to safety. If the size of the team or the blend of skills is not adequate then the risk of loss could be increased. This is a matter of judgement for the policy team but the legal requirements must be met.

The safety team has the responsibility for driving the safety message throughout the turnaround, monitoring safety performance and giving advice

1. Task step	2. Associated hazard	3. Potential loss	4. Elected precaution

Figure 10.4 Task hazard analysis proforma

and counsel on all matters relating to safety. For a large hazardous event, the company would need to decide which of the following functions it chooses to employ:

Safety officer: a senior person with safety qualifications or safety experience gained over a long period of time. Appointed some months before the event. Translates the objectives of the policy team into a safety policy and writes the safety plan for the event.

Assistant safety officer: normally a supervisor who has experience in safety matters and has the ability to administer the team. Appointed two to four weeks before the event. Takes care of the day-to-day running of the team during the event and is on hand to advise on safety.

Safety representatives: selected from workers who have a specific interest in safety. A mixture of plant and contractor personnel. Appointed a few days before the event and start working on Day 1.

Contractor safety officers: provided by the main contractors and any other contractor who has enough men on site to warrant such a presence. On a large event may be part of the contractor preparation team. If the alliance between the contractor and client is a close one, may act as turnaround safety officer on the client's behalf.

Turnaround emergency controller: selected, on the day, from plant based personnel who understand the operation of the plant emergency and toxic procedures. Responsible for taking control during an emergency. Normally has other duties.

Check list of responsibilities

Safety officer

- manage safety team and report to the turnaround manager via daily turnaround control meetings
- formulate the turnaround safety plan
- organize all turnaround safety inductions
- drive safety campaigns and initiatives
- organize daily site safety inspections
- organize random 'spot-cheeks'
- publish daily safety reports
- organize a safety suggestion/complaints book and analyse it on a daily basis
- make recommendations for changes to safety policy/procedures to improve safety

Assistant safety officer

- supervise safety team on a day to day basis
- report to safety officer
- monitor safety representatives performance daily
- monitor emergency team's performance daily
- provide safety advice to all personnel
- respond to any unsatisfactory safety condition, accident, incident or emergency
- solve first line safety problems and refer those outside scope to the safety officer
- keep all safety records and statistics
- provide information for daily safety report
- assist safety officer with safety inductions

Safety representatives

- represent the entire work force
- constantly monitor work areas to drive safety
- inspect, scaffold, platforms, accesses, barriers
- control housekeeping and general plant state
- monitor wearing of safety clothing
- monitor traffic movement on site
- monitor work practices
- record all safety infringements
- check entry guardians are in place
- take part in 'spot-checks'
- provide safety advice to all personnel

Contractor safety officers

- ensure contractor employees conform to the safety requirements agreed in the contract
- ensure all contractor employees receive safety induction and are issued with a security pass
- ensure no contractor employee is required to carry out any task he deems unsafe
- record and report any safety infringements
- attend the daily turnaround safety meeting
- if required, attend the daily control meeting

Turnaround emergency controller

- manage the turnaround emergency system
- manage the area controllers and marshals
- establish links with emergency services
- monitor day to day running of the emergency system run by area controllers and marshals
- manage all fire, toxic and incident alerts
- report on a daily basis to the safety officer
- carry out random checks on toxic refuges
- carry out random spot checks on fire equipment and assembly points
- must be contactable at all times while on site
- must deputize an area controller to cover at all other times – including night shift

Area emergency controllers

- manage the emergency system in a specific area
- co-ordinate and monitor the daily performance of the emergency marshals
- collate the daily 'refuge sheets' and report results to turnaround emergency controller
- ensure that all emergency marshals have the required safety and emergency equipment
- organize emergency marshals during a site emergency (fire, toxic or major incident)
- report to turnaround emergency controller on a daily basis
- must be contactable during the shift
- must have a deputy to cover for absences

Emergency marshal

- responsible for people accommodated in cabins or rooms designated as 'emergency refuges'
- ensure all personnel sign in and out on the emergency register on a daily basis
- call the roll in the cabin during a toxic or other emergency requiring men to seek refuge
- call the roll at assembly point during fires
- report roll calls to area emergency controller who will contact with missing person list
- search for missing persons during emergency
- lead first line fire fighting teams
- must be contactable during the shift
- must have a deputy to cover for all absences

Entry guardian controller

- manage the people selected to act as guardians for work teams entering enclosed spaces
- ensure the appropriate persons are properly equipped and are in contact at all times – contact them regularly
- respond to any distress call from the guardian and inform the assistant safety officer
- visit the guardians regularly to ensure they are performing their duties and not being harassed
- if an accident occurs, inform the safety officer, turnaround manager and emergency services
- at times of accident do not leave post to assist – be the vital communication link

Figure 10.5 The safety team

Area emergency controller: selected, on the day, from engineers or superintendents (client's or contractor's) to be responsible for the men on their areas during an emergency when they will work under the direction of the turnaround emergency controller. Have their normal duties to perform.

Emergency marshals: selected, on the day, from trade supervisors with breathing apparatus training. They will assist the emergency controllers to marshal people during an emergency and form the search and rescue teams if anyone is unaccounted for.

Entry guardian controller: selected, a few days before the event, from the safety representatives. Responsible for the personnel who act as guardians for the protection of workers covered by entry permits.

As well as the specific responsibilities of team members, as explained in Figure 10.5, there are a number of other matters concerning the team, namely:

Safety cabin: the team is housed in a safety cabin. On many turnarounds this is painted in distinctive colours (i.e. green and white stripes).

Safety log book: in the cabin there is a log book in which any person on the turnaround is free to report unsafe acts and conditions or make suggestions for improving safety.

Newsletter: it is common practice on turnarounds for the safety team to publish a daily safety newsletter to communicate all matters relevant to safety to the widest possible audience.

Safety incentive schemes: a rather vexed subject. Some companies believe it to be a good way of focusing the workers' minds on safety, while others contend that it is counter productive (for example, if awards are made for accident free days it may encourage an individual to hide the fact that he has been injured in order not to 'spoil' his colleagues' chances of winning the award). Other companies insist that safety is a statutory requirement, and therefore people should not be rewarded simply for doing what they are supposed to do anyway. The company culture determines whether or not safety awards are made and, if so, what form they will take. If incentive schemes are adopted, it is normally the responsibility of the safety team to administer them.

So far it is the preparation for safety that has been discussed, and the systems and routines put in place to minimize loss. The other, equally important, aspect of the safe system of work is monitoring safety performance.

Safety inspections (see Figure 10.6)

On turnarounds, the monitoring of safety performance is achieved by the practice of two types of formal safety inspection: daily, of the general area, and spot checks on specific jobs.

Check list of activities

The daily safety inspection

The team follow a checklist such as the one below

Daily safety theme
- is everyone aware of the theme. If not, why not?
- are supervisors enforcing the safety theme?
- are workmen complying with the safety theme?
- is anything making the theme difficult to enforce?

Unsafe acts
Is there a risk that any worker will endanger himself or others because he is working:

- with improper motivation?
- in a dangerous location/prohibited area?
- in dangerous conditions?
- with dangerous substances?
- with dangerous equipment (or misusing, abusing tools or equipment)?
- with an unapproved job method?
- without proper PPE/protection/precautions?
- without proper supervision, or alone?
- in a confined space without a guardian?
- in any other dangerous way?

Unsafe conditions
- is any permanent or temporary structure unsafe?
- are there holes/gaps on platforms or handrails?
- are there any unsafe or damaged ladders, small platforms, stairways or zip-up scaffold?
- is any live pressurized or high temperature or moving plant inadequately guarded?
- is radiography taking place without proper isolation/barrier/sign – above/below/adjacent?
- are vehicles being driven recklessly or onto areas where they are prohibited?
- are any roads, paths, walkways or stairways blocked by equipment or materials?
- is any dangerous projection unguarded?
- are any teams working inside vessels prone to poisonous, pyrophoric or electrostatic build-up?
- is there any precaution which protects plant but places people in danger?
- is there any other unsafe condition?

Housekeeping
- is the workplace dirty, oily, greasy or wet?
- is there debris or rubbish lying about?
- are rubbish skips inadequate or hidden?
- are there insufficient plastic rubbish bags?
- are nuts, bolts, washers etc. being left lying on platforms or on the ground?
- are supervisors not enforcing good housekeeping?

The spot-check

The purpose of the spot check is to ensure work teams are working safely and in safety

Spot check technique
- list a number of tasks for consideration
- select one task at random from the list
- study the task specification sheet and related documents to familiarize the team with the task
- go to the process cabin and inspect the master copy of the permit to work (photocopy it)
- visit the work site and observe work conditions and team behaviour
- announce a 'spot-check' and stop the job safely
- discuss the permit to work, the task and the working conditions with work team members
- restart the job when it is safe to do so
- record and report all safety infringements to the safety officer and turnaround manager

At the end of the spot check the team should be able to answer the following questions:

- is the area engineer managing his area properly?
- is the co-ordinator providing all that is required to ensure job safety?
- is the supervisor in control of the job?
- has the work team been properly briefed on the task and on safety requirements?
- do individuals know what to do in an emergency (fire/toxic/accident/incident)?
- is the permit to work on at the job site identical to the one in the process office. If not, why not?
- is the permit adequate?
- do the work team understand the permit, the hazards and the precautions?
- are the team complying with precautions?
- is the task specification safe?
- are the team complying with the task specification?
- are the team using the correct tools and equipment and are they using them safely?
- are individuals wearing the correct safety gear?
- are all access, egresses and platforms clear and properly maintained (check scaff-tag)?
- is there any atmospheric condition which could endanger the health of the team?
- are there any substances which could endanger the health of the team?
- is there any condition in the work area which could endanger the safety of the team?
- is there any other safety issue worth recording?

Figure 10.6 Safety inspections

The daily safety Inspection (see Figure 10.6)

The daily inspection is carried out by a three-man team drawn from a list of directors and managers, who have been invited from all parts of the company and have agreed to take part in a rota for carrying out such a duty – on dates specified on a programme drawn up by the safety officer. It is a given in any safety programme that the commitment of directors and senior management is crucial to its success. By participating in this way, the senior staff of the company are seen to demonstrate their commitment.

The duration of the inspection is normally one to two hours. The managers meet the safety officer beforehand, who briefs them on what is expected of the team. The inspection should have a daily theme (e.g. the wearing of light eye protection) as well as carrying out the general search for unsafe acts and conditions. When it is completed, the team is de-briefed by the safety officer and this generates a report concerning any occurrence of unsafe acts or conditions that the team may have found. The safety officer relays this information to the area engineers who are responsible for eliminating the unsafe acts or conditions. The safety team regularly monitors the areas to ensure this is done.

The spot check

Spot checks (sometimes referred to as 'job freezes') are carried out on a random basis by a team selected by the safety officer. They are usually of one to two hours' duration. The team meets for briefing beforehand and for de-briefing afterwards in the same manner as for the daily inspection. However, the focus is the opposite of the daily inspection, in that it is aimed at one specific task. The check tests whether the people working on the task are aware of the safety requirements and are conforming to them. The right-hand column of Figure 10.6 details the protocol of the spot check.

Unfortunately, even with all of the safety systems in place, accidents or incidents which have to be investigated may still occur.

Investigating accidents (see Figure 10.7)

Defining the term 'accident'

An accident is an event which causes death or injury to people, damage to property or pollution of the environment. Accidents can be minor, serious, major or catastrophic.

Why investigate?

Accident investigations are carried out:

- because it is the responsible and caring thing to do;
- because it is required by law;
- to eliminate the recurrence of similar accidents;

Inquiry attendees

The inquiry will involve some, or all, of the following:

Members
- Turnaround manager or nominee (Chairman)
- Turnaround safety officer (Secretary)
- Plant safety manager (if applicable)
- Plant manager or nominee
- Process supervisor
- Maintenance supervisor
- Selected safety representatives

Witnesses
- Turnaround area supervisor involved
- Persons involved (if available) and witnesses

Others
- Management or trade union observers
- Legal representatives

Establishing the facts

Reports are read, documents studied and people questioned to establish the facts of the accident and in some cases it may be necessary for the team to visit the site of the accident to establish:

1. *Loss*
- how many people were killed or injured?
- the nature of the injuries of each
- the damage to plant and property
- the nature of pollution of the environment

2. *Incident*
- what object or substance caused the injury in terms of impact, cutting, abrading, crushing etc.?

3. *Immediate causes*
- what made the incident occur in terms of defective equipment, conditions or acts?

4. *Basic causes*
- why did the immediate cause arise, in terms of lack of control, lack of awareness or incorrect motivation?

5. *Loss of management control*
- why was the basic cause allowed to arise, in terms of, inadequate systems, standards, procedures, training or communication?

6. *Recommendations*
- what can be done to regain management control, eliminate basic causes and prevent recurrence in terms of:
- improving systems/standards/procedures?
- briefing/retraining personnel?
- upgrading/improving equipment and plant?
- providing protection or warning signs?

Preparations

To prepare for the inquiry any or all of the following should be obtained:

- copy of the supervisor's first line accident investigation and report
- both copies of the permit to work
- the task specification being used at the time
- any supporting documents such as drawings, weld, inspection or pressure test specifications
- P and ID's, line diagrams and plant drawings
- specifications of all materials and substances
- specifications of tools and equipment used
- any history of similar accidents on the plant
- the names and designations of people involved
- the names of any witnesses
- the facts or an assessment of the loss
- any written report/testimony about the accident

Analysing the inadequacies

Questions to uncover basic causes

Environment
- was the work area safe, in terms of access and egress, atmosphere, configuration?

Procedures
- was the task properly organized, in terms of supervision, instruction, communications task specification, other information?

Equipment
- was equipment adequate, in terms of ergonomics, type, installation, serviceability or protection?

People
- were the people involved competent, in terms of physical and mental aptitude, attitude, training and experience, skill, motivation or stress?

Inquiry report

The report should contain the following information:

1. A summary of the accident/incident
2. Details of injuries or fatalities
3. Details of damage to property
4. Details of environmental pollution
5. The basic causes of the accident/incident
6. Recommendations to prevent recurrence of the accident/incident

Figure 10.7 Investigating accidents

- to reduce the anxiety of workers who are emotionally affected;
- to understand the root causes of accidents;
- to provide hard data for feedback to the safety system;
- to provide information to a higher authority.

If an accident does occur, an inquiry is convened by the turnaround manager to establish the facts about the five links in the accident chain, namely:

1. *loss* – the type and magnitude of loss in terms of death or injury, damage and environmental pollution;
2. *incident* – the specific incident that caused the loss, in terms of contact with a substance, object or energy source;
3. *immediate causes* – of the incident in terms of human acts and working conditions;
4. *basic causes* – underlying the immediate causes in terms of human factors and job factors;
5. *loss of management control* – in terms of systems, standards and procedures.

After many years of accident investigation in many industries in many countries, two basic rules concerning accidents have become evident:

1. The specific circumstances surrounding accidents are almost never repeated.
2. The underlying causes of accidents are always the same and are any combination of the following:
   – uneducated, incompetent or uncaring management;
   – inadequate safety systems, standards and procedures;
   – bad planning and preparation;
   – inadequate individual skill;
   – inadequate safety awareness;
   – incorrect motivation.

Safety is the business of everyone connected with the overhaul, but it is the specific responsibility of the turnaround manager to ensure that there is a safe system of work in place.

Case study

During a turnaround on an acid plant a large vessel with a dished bottom end, four-metres in diameter, was opened up for inspection. The first person into the vessel – a scaffolder (stager), whose task was to erect a platform for the inspector to stand on – slipped down the dished end and broke his wrist.

The accident enquiry investigated and found the following.
The dished end was made from highly polished (and very expensive) chrome–molybdenum steel. It had a number of stepping bars welded onto the surface to allow a person to climb down from the manhole to the bottom of the dished end but the bars stopped one metre from the bottom of the dished end so the person would have to step out onto the curved surface.

During previous overhauls the scaffolders had damaged the surface by walking on it with rubber boots, possibly contaminated with mud. The then plant manager (five years before) had written a site instruction – which was actually written on the permit-to-work! – that anyone entering the vessel must wear cloth or paper overshoes. He had rightly been concerned about the damage being done to the vessel surface but he had not considered the effect of someone trying to walk on a curved, highly polished surface without anything to hold on to.

In the previous turnaround (three years before) an identical accident had occurred.

There had been an enquiry which recommended the use of a safety harness coupled to a hand-operated hoist so that the scaffolder could descend under control. The recommendation had never been adopted.

It was also discovered that another nearby plant owned by the same company had similar vessels and had solved the problem by issuing the scaffolders with special boots with non-slip soles that did not scuff the surface. Both the boots and the harness idea were adopted for future turnarounds.

What are the lessons to be learned here? A failure to consider the safety ramifications of an instruction set this particular accident up. A failure to adopt previous enquiry recommendations caused a repeat of the accident. Useful safety information (regarding the boots) was not circulated around the company and an avoidable accident was allowed to occur. So the lessons are: always consider the safety issues when changing or modifying anything; always adopt enquiry recommendation when it makes the job safer; circulate safety information around the company so that everyone gets to know about it.

11
The quality plan

Introduction

As emphasized in Chapter 1, a turnaround is a strategic management tool used to safeguard plant reliability. The plant has three basic functions:

- to transform material from one state to another;
- to transport material from one place to another;
- to contain the material during transformation and transportation.

All of the various items of equipment on the plant are configured to facilitate the most effective performance, in line with current knowledge, of the three functions. A turnaround is performed, as part of an overall maintenance programme, to protect these three functions. However, it is a radical intervention – many parts of the plant are taken apart, worked upon, and then put back together. There is a real risk of introducing unreliability via the very act, i.e. the turnaround, which is performed to safeguard its reliability. Its planning, preparation and execution must therefore be performed in such a manner as to ensure that the integrity of the plant is not adversely affected.

Many companies employ auditors (see Chapter 1) to examine their performance, highlight any shortcomings and recommend improvements. To assure quality, the requirements of every task must be correctly specified and then performed to that specification. Everything must be done correctly. The way to 'get it right' is to have a coherent, auditable quality trail from initial work request to final acceptance of the completed task. Figure 11.1 is an example of a task preparation checklist (indicative, and not exhaustive). By following it the planner can ensure that all the necessary steps have been carried out, and an auditor can follow the logic.

The type of quality system used for any particular turnaround, and which will be established by the policy team, will depend greatly upon the general approach to quality within the company.

Policy team issues

Figure 11.1 (Right) Sample audit trail for task preparation

Yes, No
or N/A

01. Has the work request been endorsed by the plant manager or nominee?

02. Have plant drawings/documents been validated by the plant manager or nominee?

03. Is issue of above drawings and documents controlled by the planning officer?

04. Has the plant manager specified the plant standards and operating procedures?

05. Has the work scope meeting validated the task?

06. Have plant personnel tagged the task on site with a unique number (tag number)?

07. Has turnaround manager specified if task is to be done in house or by a contractor?

08. Has the planner specified the task on a task sheet or list and tag numbered it?

09. Has the planner specified materials, with material certificates for critical materials?

10. Has the planner specified equipment with calibration certificate for measuring instruments traceable to national standards?

11. Has the planner specified all required services and utilities?

12. Has the planner cross-referenced supporting documents on the task sheet and defined them on an index sheet to create a task package?

13. Has task package been approved by maintenance/operations managers or nominees?

14. Has the planning officer scheduled the task in the turnaround plan?

15. Have all materials/equipment/services been procured by site logistics team?

16. Have materials/equipment/services been checked on arrival at site by the appropriate person, i.e. planner, engineers, inspector etc., and:

16.1 has any item been rejected/quarantined (in writing) by the 'inspector'?

16.2 has any item been concessed in writing by the plant manager?

16.3 does critical material have a material certificate?

16.4 does each measuring instrument have a current calibration certificate?

16.5 have all rejected items been returned to the supplier and re-ordered?

16.6 have all items been properly stored and protected?

16.7 is there a controlled procedure for issuing items from stores?

17. Has the planner issued a copy of the task package to the supervisor?

18. Has the planner shown the supervisor the task on site and briefed him on the task?

19. Has supervisor checked the task tag number against tag number on the task package?

20. Has the supervisor checked that material/equipment/services needed for the task are available together with any material and calibration certificates?

21. Has the supervisor ensured that personnel chosen to do the task are competent?

22. Has the supervisor briefed the personnel on the skill and safety aspects of the task?

If the above activities conform to requirements then the task may proceed

Quality = conformance to requirements   is vital to the success of any turnaround
Unfortunately, if the amount of time, effort and money which has been expended
on it over the last decade is anything to go by, its promotion does not seem to
have been an inherent characteristic of many industrial enterprises. Assurance
and control of the quality of any project require a system dedicated to that end.
Not having one for a turnaround can lead to over-run, overspend and inferior
work that will compromise plant reliability.

The policy team's approach to quality will depend greatly on the current
level of quality awareness within the company. In many cases the team will
look to the turnaround manager to guide them on the question of quality
systems. He must investigate the company's existing systems and, depending
upon what is found, take one of the courses of action that will now be discussed.

ISO 9002

If the company operates an internationally or nationally accredited quality
system such as ISO 9002, the turnaround manager should adopt it.

Internal quality system

If the company has a well defined set of internal quality procedures, the
turnaround manager should adopt them.

Fragmented procedures

If the company has a number of procedures which cover quality but are not
integrated into a system, they should be integrated into one system and,
where necessary, augmented by writing additional procedures.

No formal quality system

If no formally written down procedures are followed, the turnaround manager
should ascertain if there is an informal system at work and, if there is, whether
or not it is adequate. If it is, it should be formalized by writing simple procedures
and, if necessary, augmented by writing additional procedures.

No system

If the company has no formal system and the informal one is inadequate, the
turnaround manager should write out a set of simple quality procedures and
check-sheets to cover the turnaround.

Whichever of the above options is chosen, the turnaround manager should
propose the quality plan for the event to the policy team, who should then
discuss the requirements and come to an agreement on what the approach to
quality will be. It is the turnaround manager's responsibility to propose the
quality plan, but it is the responsibility of the policy team to approve it. They
must also bear the consequences if there is a failure due to the lack of an
adequate quality system.

Basic quality requirements

A turnaround is a human activity system which requires all of the elements shown below to work effectively
If any element is missing or inferior, the system will malfunction and fail to deliver the agreed outputs

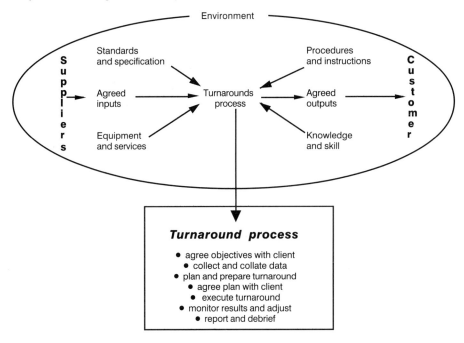

Figure 11.2 Basic quality model

A turnaround is a human activity system, that requires all of the elements shown in Figure 11.2, and detailed in Figure 11.3, to be in place and working effectively. If any element is missing or inadequate, or if the environment within which the event is being performed is hostile, the system will malfunction and fail to deliver desired outputs. It is not within the scope of this book to delve into the mechanisms of quality assurance and control. The emphasis here is on how the need for a quality approach impacts upon the turnaround manager. There are many excellent books on the subject of quality and these should be consulted if a deeper understanding of the subject is required.

Hold points

For critical items of equipment, the job methods should be analysed to find the activities which require to be monitored by the client's inspectorate to assure that the quality of the work being performed is satisfactory. Hold points – at which work must be inspected and signed-for by the client before the next activity can commence – are then inserted into the job sequences. These points usually occur after alignment, setting,

Suppliers

- include makers/vendors/distributors/contractors
- state the requirements specifically to the supplier
- negotiate and agree requirements with the supplier
- make the agreement formal and written
- ensure the suppliers are capable of meeting the requirements
- alert the supplier if they do not meet requirements
- effective feedback is essential

Customer

- the person or agency who has to pay for, and live with, the output of the turnaround process
- without a customer there is no need for the process
- agree requirements with the customer
- ensure the organization is capable of meeting agreed requirements
- always strive to meet agreed requirements
- alert the customer to anything which will prevent meeting requirements
- effective feedback is essential

Inputs

- anything that will be transformed by the process
- includes data, raw materials and proprietary items
- during the initiation stage the customer supplies data – the normal relationship is reversed
- ensure there is a specification for inputs
- check inputs to ensure they meet requirements
- remember the GIGO rule; garbage in – garbage out
- once in possession of physical inputs, do not expose them to damage or deterioration

Outputs

- what is presented to the customer when the process is complete – must meet requirements
- the final output is the accumulation of every individual output – it should add value
- there are four types of output, viz.:
  1. predicted/desirable – *the planned outputs*
  2. predicted/undesirable – *effluent and waste*
  3. unpredicted/desirable – *serendipitous events*
  4. unpredicted/undesirable – *counter-intuitive failure*

Standards and specifications

- written documents that define the maximum and/or minimum tolerances on quantity, quality, time, cost and safety
- may be agreed, self imposed, externally imposed or legislated
- investigate failures to meet standards
- if outputs continuously fail to meet standards, check the standards – they may be unattainable
- ensure everyone involved is briefed on standards that are relevant to them
- remember to feedback

Procedures and instructions

- procedures are documents that define what activities have to be done, when, where and by whom, to perform the work process
- instructions are documents that define how the activities should be performed
- amend either document if a better method is found
- regularly review and renew both documents
- control the issue of the documents
- ensure everyone involved is trained to use the procedures and instructions relevant to them

Equipment and services

- equipment is any artefact used to transform inputs
- services consist of:
  - buildings and associated activities (catering etc.)
  - specialist contractors (water washing etc.)
  - utilities – gas, water and electricity
- equipment and services must be appropriate to the task, serviceable and well maintained
- alert suppliers if equipment and services are substandard – effective feedback is essential

Skill and knowledge

- competent resources must be available
- competence depends upon attitude, aptitude, training and practice to gain effective experience
- 80% of the things which adversely affect levels of competence are the fault of bad management
- the workers must be capable of doing the work
- supervisors must be capable of leading the men
- engineers must be capable of controlling their turnaround areas
- managers must be capable of running the system

Turnaround process

- appreciate the customer's needs
- validate all collected data
- plan – the right thing/the right place/the right time
- prepare – anticipate things that may upset the plan and eliminate them or protect the plan against them
- make effective decisions on what actions to take and then follow the actions through to the end
- execute the turnaround cost effectively
- monitor outputs and adjust if necessary
- report and debrief the team and the customer

The environment

- context within which the turnaround is performed, legislation, competitors, weather etc.
- cannot be controlled but may be influenced
- will be influencing, and sometimes controlling
- changes without warning – prepare against this
- can cause sudden failure of the process – a contingency plan must be prepared for this
- is not 'part of your business', but if ignored it may destroy the business

Figure 11.3 Basic quality requirements

radiography and pressure testing activities and before 'boxing up' activities.

Joints

The breaking and making of a joint is, on the face of it, one of the most straightforward of tasks on a turnaround, but it is also the most frequent one; not uncommonly, many thousands of such jobs will be required. In addition, joints are critical to containment function and badly made joints are one of the fundamental causes of duration over-runs (not to mention the potential hazards caused by uncontrolled emissions). All leaks should be deemed unacceptable but there are some joints which are more critical than others and they, at least, should be subjected to a quality inspection to eliminate, as far as possible, the possibility of leakage on start up. Figure 11.4 is an example of a joint check sheet. At the very least, the client should specify a list of critical joints which have to be inspected before making.

Case study

Some years ago the author was involved in a turnaround on a plant which suffered from a large number of leaks on start up. It had to be brought off line three times, and four days' production was lost due to this. After the event a corrective action team was formed, as part of a quality improvement programme, to try to identify the root cause of the leaks.

The team was made up of mechanical fitters and supervisors and chaired by the author. A brainstorming session was organized to generate as many reasons as possible which might account for a set of two related joints failing by leakage. No less than sixty three different possible factors were postulated, among which were:

- radial and axial misalignment of the two associated pipes;
- incorrect cold gap between flanges prior to tightening;
- poor flange face quality;
- flanges thinned due to previous over-machining;
- poor gasket quality;
- poor quality of joint assembly;
- employment of incorrect bolt tightening sequences;
- inadequate competence of the technician doing the job;
- poor quality of inspection or monitoring;
- the integrity of one set of joints being disturbed by the inferior assembly and tightening of another set of joints on the same pipe run (a surprising insight).

A previous case was quoted concerning a fracture in the weld neck of a pump on an LPG line in an oil refinery, which had caused a fire and had serious cost implications for the operating company. When the pump was stripped out it nearly caused another accident because the two pipes to which it was attached sprung approximately four feet (1.25 metres) axially due to

Critical Joint Inspection Routine - Stripping the joint

1.   What is the tag number and orientation of the insert (pump or valve etc.) ?

2.   Have any of the joints been filled with resin, if so, how will the resin be removed ?

3.   When the resin is removed, is there any evidence to suggest what ma have caused the leak ?

4.   Are the pipe and insert flanges the 'correct' schedule ?

5.   Is there an excessive cold gap between the pipe and insert flanges ?

6.   Does either pipe 'spring' when unbolted - due to axial misalignment ?

7.   Are the bolt holes in the two pipe flanges 'off pitch' - due to radial misalignment ?

8.   Is there - scoring, gouging or pitting on either of the flanges ?

9.   Are the flange faces correctly machined - plain, stepped, serrated etc. ?

10.   Have either of the flanges been over machined - to below acceptable wall thickness ?

11.   Is the gasket being removed from the joint the correct one ?

12.   Is the gasket damaged, indicating there is something wrong with the joint ?

13.   Are the fasteners used on the joint of the correct type, and are they undamaged ?

14.   Are the fasteners to be reused, if so, how will they be protected until required ?

15.   How are the two 'open' pipe ends to be temporarily sealed until the joint is reassembled ?

The results of the above survey should be reported to the Area Engineer so that action taken to eliminate any faults.

Critical Joint Inspection Routine - *Remaking the joint*

1.   Was the joint inspected when it was stripped - if not, why not ?

2.   Have the necessary refurbishment and machining actions been performed correctly ?

3.   Have either of the flanges sustained any damage since the joint was broken ?

4.   Has the correct insert been replaced in the correct orientation ?

5.   Have the two mating pipe flanges been aligned (axially and radially), without the use of force ?

6.   Are the gaskets correct and free from faults ?

7.   Have the insert and the gaskets been reassembled in position, without the use of force ?

8.   Have the fasteners been passed as fit for purpose and are they properly lubricated ?

9.   Have the fasteners been inserted in the bolt holes, without the use of force ?

10.   Have the fasteners been tightened in the correct sequence ?

11.   If required, have the fasteners been tightened to the specified torque, using calibrated equipment ?

12.   If the joint has been 'flogged', was it performed by an appropriately experienced craftsman ?

13.   Is the joint acceptable - if not, state the reason ?

Any concerns with regard to the quality of any joint should be reported to the Area Engineer so that appropriate action can be taken.

Figure 11.4 Joint check sheet

misalignment. The amazing thing is that this pump had been changed a number of times in the past, using an informal procedure for making and breaking the joints – employing two pull lifts and tang spanners – which had been developed to cope with the misalignment! No one had ever thought to cut and re-set the pipes to eliminate the fault. The consequences of this particular piece of poor quality thinking were a fractured pump, a fire, a near-miss incident and a significant loss of profit for the company.

An interesting outcome was that the team, when asked to come up with a definition of the very common term 'flogging' (tightening joints by use of a spanner and hammer, a very common practice in the petrochemical industry) decided that the best one was 'the uncontrolled tightening of a joint'. It is significant that many more critical joints are now tightened using torque equipment and bolt tensioning than was formerly the case.

12
The communications package

Introduction

In preparing a turnaround, many thousands of pieces of information are processed into plans and schedules which are created so that the work is carried out in a particular manner. If the objectives are to be met, all involved need to understand what is required of them. Without effective communication the event can go off the rails.

Issues and consequences

A major turnaround is a complex event, the management of which exposes a number of important issues including, *inter alia*, the following:

- a large volume of work must be done within a very short time;
- the work is performed by a large number of people from many different organizations;
- many different types of work are performed at the same time, in the same place and on different levels of the plant;
- some of the work will be hazardous;
- some of the work will be unfamiliar;
- the onset of emergent work may cause priorities to change, rapidly.

A consequence of this is that the potential for accidents, conflict, error and confusion is greater than normal, so it is vital that everyone involved in the event, or who can influence its outcome, is fully and properly briefed on all its aspects and requirements. The turnaround manager, assisted by his preparation team, must therefore produce a formal document that is used to brief all personnel and which will ensure that everyone gets the same message. This is the meaning of communication prior to a turnaround.

Communication

Communication, prior to a turnaround, is carried out to provide accurate general information, to alert everyone to the rules governing the event, to create a common understanding of requirements and to gain the commitment of the people involved. In a prior briefing session, the following subjects should be covered:

- the purpose of the turnaround;
- the turnaround organization;
- key dates and events;
- working patterns;
- the turnaround work scope;
- contractors;
- cost profile;
- quality plan;
- safety issues;
- facilities;
- the turnaround objectives.

The general briefing

The general briefing, based on the previously outlined agenda, is delivered to all groups who have a stake in the event. The briefing programme usually starts four weeks before the start of the event. The following groups (each one typically of between 20 and 50 people) must be briefed:

- the turnaround policy team;
- business and marketing managers;
- the turnaround control team;
- plant managers, engineers, supervisors and workers;
- local company resource groups (mechanical, electrical etc.);
- all contractors;
- project teams;
- inspectors;
- workshop personnel;
- all support groups (engineering, technology, process etc.);
- all emergency services (both company and community);
- local authorities and other external groups likely to be impacted by the event.

Timing

Obviously, everyone cannot be briefed at once so a briefing programme is arranged. The timings shown below are indicative:

Policy team	–	four weeks before the event
Business management	–	three weeks beforehand
Control team	–	two weeks beforehand
Plant personnel	–	one week beforehand
All other local groups	–	two to three weeks beforehand
Contractor labour	–	the day they arrive on site
Casual visitors	–	require safety briefing only

Delivery

Presentation of the briefing is carried out by a small group of key people, viz.:

- The turnaround manager (who presents the turnaround issues).
- The plant manager (who presents the plant-based issues).
- The maintenance, or engineering, manager (the engineering issues).
- The safety officer (the safety issues).

In this way the respective elements are delivered by the person best qualified to do so.

Format

The briefing is presented both as a written document for circulation to key personnel and as a series of overhead transparencies to be presented at the briefing sessions. Most plants have a safety video which is shown to anyone visiting the site. If it is used as part of the briefing, it must be borne in mind that it refers to a running plant – which has one specific set of conditions and hazards – whereas the turnaround will take place on a plant that has been taken off line and opened up. This will present a totally different set of circumstances and hazards. It may well be the case, therefore, that the plant safety video is of little or no use in this context. The decision to use it or not should be taken jointly by the turnaround and plant managers. Figure 12.1 is a check list of the topics covered by the general briefing.

The major task briefing

Major tasks are those that are large, complex, of long duration or unfamiliar. They will have been planned by a team under the leadership of the preparations engineer and then vetted by the plant engineer in charge of the area in which the work will be executed. With the best will in the world, the engineers responsible may miss a vital element, or remain ignorant of an important piece of information, or of the job method. While being adequate from an engineering point of view, the omission of the detail may pose other problems, not the least of which may be those impacting on safety.

Because these jobs are deemed to be so critical to the success of the turnaround, they should be subjected to a final validation process, known as the *major task briefing*, during which the area engineers must present the specification and methodology for each selected task to an open forum of managers and engineers, to test their validity. The turnaround manager arranges the meeting and selects the significant major tasks for each area. The area engineers prepare a detailed presentation on the intended execution of each of the selected tasks.

Topics to be considered

1. Purpose
- perform statutory inspections
- repair/replace faulty equipment
- investigate process difficulties
- modify existing plant
- install new plant
- introduce new technology
- demolition of redundant plant

2. Organization
- who is who/where they fit in
- control team members and roles
- size and shape of organization
- different companies/ departments involved
- different teams/their functions
- different areas/unique features

3. Key dates and events
- date plant comes off line
- date plant ready for turnaround
- date of significant or unusual events
- date of any interruption to work
- date of turnaround completion
- date plant comes back on line

4. Working patterns
- normal working day
- normal overtime patterns
- extended overtime requirements
- shift patterns – 1, 2 or 3 shifts
- 8 or 12 hour shifts
- 24 hour coverage
- time off in lieu
- which resources will work which overtime or shift patterns

5. Work scope
- capital projects
- modifications
- construction work
- demolition work
- plant overhaul:
  - major tasks (critical paths)
  - small tasks
  - bulk work
- unusual or hazardous tasks

6. Contractors
- how many contractors
- contractor company names
- which contractor does what
- specialist contractors
- interface between contractors
- interface between contractors and local resources

7. Cost profile
- overall cost of the event
- cost breakdown by area
- cost breakdown by:
  - management and planning
  - local and contract labour
  - materials
  - tools and equipment
  - services, utilities and facilities
- cost of major tasks
- cost of projects and modifications
- any other significant costs

8. Quality plan
- approved quality procedure for the turnaround
- quality auditing (if required)
- plant quality team and function
- quality objectives:
  - only approved material and equipment used
  - all instruments calibrated
  - all work completed to spec
  - critical joints recorded
  - all joints leak free

9. Safety issues
- permit to work system
- personal protection equipment
- safety and emergency equipment
- safety rules and procedures
- emergency procedures
- double checking of tagging
- isolation practice and control
- working near live systems
- working with electrical equipment
- observing barriers and signs
- personal safety issues
- personal monitoring
- general housekeeping
- hazardous substances
- safety team and safety cabin
- safety complaints and suggestions
- safety initiatives
- safety targets (drive them home!):
  - zero accidents, incidents
  - zero health hazard
  - zero environmental pollution

10. Facilities
(use plot plan to indicate)
- site security and control
- vehicle car parks and restrictions
- daily accommodation
- canteens and mess halls
- washrooms and toilets
- emergency facilities
- first aid facilities
- ambulance and hospitals
- out of hours facilities

11. Objectives
- to meet all targets set for:
  - duration
  - cost
  - quality
  - safety
- to complete all scheduled work to technical specification
- to hand the plant back in a condition that will allow it to run leak free till the next shutdown

Figure 12.1 The general briefing

The forum

The personnel who are invited to the major task briefing are drawn from the following:

- the turnaround policy team;
- nominated plant and engineering staff;
- safety manager and officers;
- managers of support groups.

The format

The turnaround manager gives an overview of the major tasks selected and highlights any significant features, such as:

- the ultimate critical path task and the critical path task in each area;
- any special techniques being used for the first time;
- any new technologies being introduced;
- degrees of difficulty of the work to be performed.

Following this introduction, each area engineer in turn presents a detailed description of the selected tasks, using the following format:

- the work to be carried out;
- the planned order of work;
- the techniques and technologies to be used;
- numbers and types of labour to be used;
- any assumptions being made;
- any possible difficulties and the strategy for overcoming them;
- any possible contingencies and allowances for them;
- any known hazards and strategies for eliminating or guarding against them;
- a rescue plan. (if required);
- overall cost of the task and specific significant costs.

Members of the forum should be encouraged to ask questions, challenge assumptions and make suggestions during the presentation in order to test its validity. The outcome of this session may be that either the presentation is accepted by the forum as it stands or that modifications are imposed.

The major task briefing should be held about four weeks before the event, to allow time for any proposed modification to be carried out.

Other briefings

Several additional briefings are required prior to the start of the event. They are the responsibility of:

Area engineers   – who brief their teams on the area work scope, area objectives, potential problems, roles and responsibilities;

Site logistics officer	– who briefs contractors on their specific responsibilities, as well as site discipline and security;
Project managers	– who brief plant and turnaround control teams on the scope of their projects, and how they will interact with the turnaround schedule;
Plant manager	– who will brief the process, Permit to Work and quality teams on the standards of performance required;
Engineering/maintenance manager	– who will brief the maintenance and quality teams on the standards of performance required;
Contractor safety officers	– who will brief their employees on the site safety rules and reinforce the safety messages presented at the general briefing.

Effective communication is vital to the success of the event and the turnaround manager is responsible for ensuring that this is achieved prior to its start.

13
Executing the turnaround

Introduction

It is impossible to convey, by means of the written word, the experience of executing a turnaround, especially a major one. The event is the realization of the plan and its course will be affected by how nearly that plan matches the reality of the event. Tens of thousands of individual activities will have been interlocked in an intricate plan – and it only takes one unforeseen or unforeseeable event to trigger the plan's unravelling.

Managing a turnaround will provoke the whole range of emotions, from the buzz of excitement which comes from successfully juggling so many balls in the air at one time right through to the pit of despair (usually experienced, in solitude, in the small hours) when the manager finally has to admit, given the current circumstances and constraints, that one or more of the objectives will not be met.

What follows in this chapter is a set of processes and routines which serve to allow the turnaround manager to understand what is going on and thereby retain control of the event.

The event

The day finally arrives when all planning and preparation are (hopefully) complete, the materials and equipment are in place and the resources have been organized and briefed. The business of shutting the plant down and executing the turnaround begins. Although many thousands of varied activities will take place during the event, the turnaround manager should determine that from his perspective as a work manager there are, in fact, only two types of work that have to be controlled, viz. routine and unexpected.

The routine

Work that has been planned, scheduled, and resourced and, where necessary, appropriate allowance has been made for contingencies. The emphasis here should be on exercising control over this work, by means of the turnaround planning and control system, to meet or beat the schedule and budget.

The unexpected

A turnaround is a complex event during which problems will arise which have been neither predicted nor expected – no matter how good the planning and preparation. They could be caused by such things as changes of intent, accidents, industrial conflict and so on. The aim is to be prepared – to react speedily and effectively to the situation, minimizing any negative impact. If the incident is not properly controlled the routine can rapidly become the unexpected and the unexpected may become the catastrophic.

Shutting the plant down

Shutting down is controlled by the plant manager and his personnel. It should be carried out in conformance with the shutdown network plan, which, as part of the overall turnaround schedule, will have been formulated or updated by the plant team during the preparation phase. In many situations, especially those in which the plant is being shut down for the first time, or by an inexperienced team, the logic, timing and duration of events will be based only on judgement and on whatever experience is available. Reality may force the plant team to deviate from the plan; if this happens it can have two effects:

(i)   it alters the logic of the shutdown network and affects the timings and durations of activities;
(ii)   it affects the start of the mechanical duration.

The plant manager must inform the turnaround manager so that the effect of the deviation on the existing schedule can be rapidly calculated, and the schedule amended if necessary. Failure to do so may render the existing schedule useless as a control tool.

Prior to the start of the shut down phase all necessary tools, equipment, materials and isolation plates are identified and laid out on the plant so that they will be available when required. The turnaround manager supports the plant manager and supplies resources to carry out the following supporting tasks:

- transporting shutdown equipment around site (hoses, drain valves etc.);
- fitting and removing isolation plates;
- positioning and connecting water cooling and washing equipment;
- fitting shutdown valves and pumps;
- breaking and re-making joints to the plant team's instructions;
- cleaning up spillage of fluids and other substances;
- carrying out any other task which progresses the shutdown;
- the plant team 'cools' the plant down and decontaminates it before handing it over to the turnaround manager. A plant may be handed over in total or one system at a time.

Some of the typical faults which cause the shutdown to over-run and intrude into the time allowed for the mechanical duration are:

Problems with physical items – items which are necessary to shut the plant down (tools, plates etc.) are either not there when they are required or are found to be unsuitable.

Duration over-run – activities take longer than planned. This may be due to the planned time being unrealistic (it may have just been a guess) or the process used to perform the activity being ineffective (e.g. four tonnes of steam being used when 10 was required).

Emissions and spillage – product or service fluids are emitted or spilled and have to be cleaned up before work can proceed.

Inexperience – personnel employed to shut the plant down not having the necessary experience to do so in a timely and effective manner.

Disorganization – the shutdown team leader loses control of the work due to bad planning, poor briefing or ineffective communication.

The shift system

On shutdowns which take more than one shift to complete, time can be lost if the shifts have different approaches to shutting the plant down or the incoming shift insists on re-checking some of the work carried out by the outgoing one.

Tight management of the shutdown on a twenty-four hour basis by experienced shift managers who have agreed the logic of the shutdown, stick to it and perform an effective handover at the start of each shift will go a long way to eliminating these problems.

The routine (see Figure 13.1)

The turnaround manager's routine

A turnaround can typically last anything from one to six weeks. In order to remain in control, the turnaround manager must set up a daily routine which will allow him to deal with all critical items and stay in touch with what is going on. It must include personal contact with key people in the organization on a regular basis. It also does no harm (and often much good) if he talks to the workmen on the job, often gaining insights into the situation which he would otherwise miss.

To a great extent, the turnaround manager must rely upon other people to feed him the information he needs to make decisions and take actions but it would be a foolish manager who allowed that reliance to become a dependency. He must be able to verify that the information fed is in accord with the overall progress and performance of the work.

A typical daily routine for a turnaround manager will now be outlined. It may not always be possible, nor even desirable, to carry out the activities in the order shown but it is important that they are carried out with daily regularity. Everyone else on the turnaround takes their cue from the turnaround manager and if they can discern a regular pattern they will be more comfortable.

The most effective way for a turnaround manager to control the rate of progress on an event is by means of a daily routine.

The daily routine helps the manager keep his finger on the pulse of the event and apply pressure when and where it is needed.

Actions

- Check previous 24 hours progress with the planning officer
- Check cost control and forecast with the cost engineer
- Visit safety cabin and check on any safety issue
- Visit the stores and check on material delivery and issue problems
- Visit the workshops (if feasible) and check on work progress
- Tour the site to check on safety and housekeeping – talk to people!
- Regularly take part in safety inspections and spot checks
- Visit permit-to-work office and check on any problem
- Visit the quality team and discuss any quality issues
- Vet overtime requests and approve/amend/reject

Meet with the plant managers and:

- resolve any current technical problems
- discuss and approve/reject requests for emergent work
- formulate strategies to keep the event on programme
- discuss and resolve any industrial relations problems
- discuss and resolve any interface problems
- define the consequences of any change of intent

Chair the daily turnaround progress meeting:

- safety officer reports on safety issues and initiatives
- area engineers report on area work progress and issues
- project managers report on project progress and issues
- plant manager reports on any plant related issues
- maintenance manager reports on any engineering issue
- quality team leader reports on quality issues
- cost engineer reports on expenditure and cost issues
- chairman sums up/makes decisions/delegates actions

Write a daily progress and safety report, and issue it

Figure 13.1 Turnaround manager's daily routine

The process used to fulfil the purpose of the routine is:

- examine the situation to gain understanding;
- decide if there is a problem and, if there is, define it;
- expose the root cause of the problem;
- create a solution for the problem;
- take action or delegate action to implement the solution;
- record the action taken;
- monitor the effectiveness of the solution.

The process must become automatic because solutions to the many problems which arise during an event must be created on the run. There is little time for deliberation and everyone else is looking to the turnaround manager to keep things on track – he is probably the only person who ever sees the whole picture. Anyone who manages a turnaround should get plenty of opportunity to perfect this skill.

The turnaround manager's daily routine should include, but not be limited to, the following procedures.

Night shift progress

The turnaround manager should check, first thing in the morning, the progress and performance for the previous night shift. On the turnaround schedule, for the purposes of controlling resources and durations, work done on night shift is treated in the same way as work done during the day, but it has to be remembered that at night there is usually only a skeleton control crew on duty and most of the key personnel are off duty.

Normally, night shift is used to progress the critical path work and any work which has fallen behind schedule – the most crucial work on the turnaround – and yet it is often the least managed activity. It only takes one thing to go wrong, an incorrect briefing, issue of the wrong materials, an equipment breakdown, to halt work. Getting it started again can be difficult because the normal services and personnel are not available. Even call-out takes time and sometimes fails due to bad communication.

Critical path hours (which may represent 30 to 40 per cent of the total hours allotted to the job) that are lost on night shift can never be recovered. Night shift work needs to be controlled by a strong and resourceful manager and should not be used to blood inexperienced engineers. Night shift planning, preparation, resourcing and briefing must be of the highest standard.

Control of work

The turnaround control team will have generated a number of documents to control work and the manager must keep up to date with current progress by visiting the planning office (or cabin) daily to get an update on the progress of the event. He should examine and analyse documents, including, among others:

> *The turnaround schedule* – comprising the shutdown network linked to the mechanical duration schedule and the start-up network. This is updated daily. It shows the overall progress in every area and indicates whether a job has gone critical (i.e. fallen behind planned schedule).
> *Look ahead schedules* – abstracted from the turnaround schedule and show the work needing to be done over the next few days. They are especially useful to the supervisors who are executing the work on site.
> *Run-down graphs and S-curves* (see Figures 13.2 and 13.3) – showing actual and projected performance against planned performance. The variable plotted can be cash expenditure, or completed programme-hours, or expended man-hours, etc.
> *Work control sheets* – list batches of similar jobs (vessel inspections, scaffold

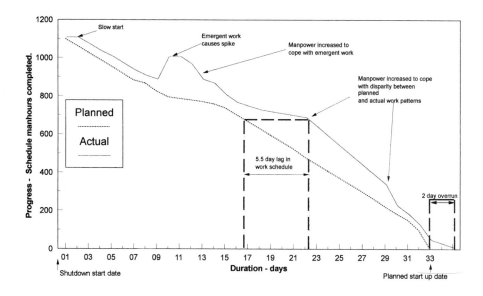

Figure 13.2 Run-down graph: Area X

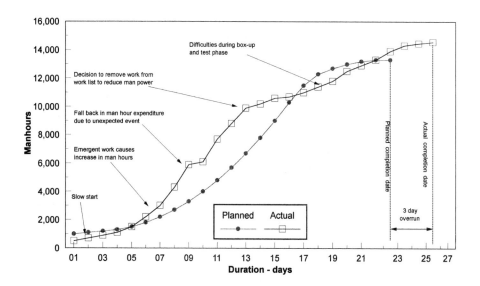

Figure 13.3 Example of an S-curve: Expended manhours

builds etc.) and show the key stages of each job where, as part of the single point responsibility principle discussed in Chapter 7, signatures are required to indicate that that particular stage has been completed.

These are all visual aids which allow the manager to see, at a glance, the state of progress overall and on all the individual areas.

The validity of the schedules and control documents depends, however, upon their accurate updating. If this is ignored a false picture of progress can be created. There is a practice among some supervisors of 'keeping some hours in the back pocket'. This means that they fail to reveal all of the progress that has been made – especially if they are ahead of programme on a particular job – because they know from experience that problems will occur as they go through the event and if they are held up they can feed some of the 'back pocket' hours into the record and appear to be working on schedule. This practice distorts, propelling into the critical zone work which has already been completed but not reported. The schedule may even indicate to the turnaround manager that he is behind schedule when he is not – prompting him to take measures to respond to a situation which does not exist. The manager who does not have his ear to the ground may be stunned by the apparently huge amount of work which is completed on the final day as supervisors dump all of their remaining back pocket hours into the schedule. The work is completed on time maybe, but the manager may already have expended time, resources and money to recover from the fictitious backlog.

The other side of the coin is represented by serious problems with performance or progress which are not reported by the area teams – lulling the manager into a false sense of security. The turnaround manager may believe he is on programme when, in fact, he is well behind. Again, but with more serious consequences, he does not find out until it is too late to do anything about it. The turnaround manager must know the true situation on the event if he is to exercise control and this knowledge only comes from constantly monitoring and questioning the updating information.

Critical path job (CPJ)

Although duration is associated closely with control of work, its importance merits a paragraph or two to itself. At all times, the turnaround manager must be aware of a number of factors which have a direct bearing on the critical path (which will determine the actual duration of the turnaround).

Most other types of problem on an event may be resolved within the planned end date by the judicious use of overtime, extra resources etc. This is unfortunately not true of a job which lies on the critical path; programme time lost there can normally only be recovered by taking work out of the CPJ. However, this is only very rarely possible at any time during the event and virtually impossible late on in the event.

To ensure that the state of the critical path is known, there are several questions which the turnaround manager should ask on a daily basis, viz.:

1.   Which job is currently the CPJ and what type of work does it entail?
2.   Is the CPJ on, ahead of, or behind schedule? If behind – why, and what can be done about it?
3.   Are there any known issues that will cause problems on the CPJ?
4.   Is there anything that will cause a job not currently on the critical path to become the CPJ, if so, can anything be done to avoid this?

The turnaround manager and his control team should devote some time every day to an analysis of the current situation on the critical path, in order to keep it under control as far as possible. The worst case scenario – which reflects, as nothing else can do, the uncertainty at the heart of a turnaround – is the emergence of a significant piece of extra work on the critical path job (e.g. damage is found, or a pressure test fails on a welded butt and weld repairs are required), which adds to the duration of the event, and nothing can be done about it. In this case, the turnaround manager must involve the policy team who must analyse the effects and publish a new duration for the event.

Expenditure

The cost engineer should produce a daily report detailing current expenditure (actual and committed) and AFC (anticipated final cost) for each of the designated areas of the turnaround as well as for the overall total. This information allows the turnaround manager to analyse expenditure patterns and pin-point activities which are overspending. This can occur unexpectedly in straightforward activities such as scaffolding and lagging, rather than in the more highly technical activities where it might be anticipated, but which are often more tightly planned and controlled. The manager is able to expose root causes and propose remedial action. Also, the report is backed up by more detailed information which can be used as necessary. This is especially important if a large amount of emergent work is driving expenditure upwards. Copies of the report should be sent to the members of the policy team to keep them in touch with the situation because, if the AFC increases beyond the forecasted budget (based on the final estimate) the two main responses are to find more money to cover the extra work or to prune the work list – either way the policy team must be involved. Prior warning is preferable to an unpleasant surprise when it is too late to respond (e.g. finding out that the event has overspent after it is finished).

Safety

The turnaround manager should visit the safety cabin and check on any current safety issues. *Every day* – there is always news on safety and it's usually bad. It is one thing to formulate a safe system of work during the relatively calm days of the preparation phase when everyone is thinking rationally. It is quite another to try to enforce it on hundreds (in some cases thousands) of people who are working under time pressure and (as has been proved so often in the past) will act irrationally and thereby put the health and lives of themselves and others at risk in order to progress the job.

Even more irrational seems the almost obstinate refusal of some people to wear the protective equipment they are issued with and obey the rules which have been set up purely for their protection. Housekeeping is a very mundane subject which is often ignored and yet bad housekeeping generates unsafe conditions which are a major contributory factor to accidents.

The turnaround manager has ultimate responsibility for the safety performance of the event. He must exercise constant vigilance and monitor the safety situation regularly in order to ensure that the safe system of work is operated to the full.

Site logistics

The site logistics cabin should be visited and the current situation and any anticipated difficulties noted. The current disposition of tools, equipment, materials, proprietary items, consumables, services, utilities, accommodation and facilities must be understood and any problems, requiring the input of the manager, resolved.

Items for special consideration should be:

- disposal of toxic or effluent substances;
- violations of site rules (such as use of restricted areas for storage etc.);
- utilization of cranage, heavy plant and vehicles;
- any issue regarding accommodation or facilities which is likely to cause industrial relations problems (a shortage of toilet paper can be as troublesome as a shortage of materials!).

Workshops

The turnaround manager (or his nominee) should regularly visit workshops to check on progress. Nowadays, most workshops are off-site and it is imperative that the old adage 'out of site, out of mind' does not apply. If progress is not monitored not only can work fall behind schedule but the workshop may be working to the wrong priority. The items being worked on must be delivered back to site before or on their due dates. A check should also be made on the state of items being delivered to the workshop to ensure that any required on-site decontamination has been carried out. If marshals are being employed to organize the movement of bulk items between site and workshops they should be included in the regular workshop visits.

Site safety

Monitoring is a vital component of the safe system of work. Using the unsafe acts/unsafe conditions routine, the manager should tour the site daily to check on safety and housekeeping. This must include talking to as many people as time allows and constantly pressing home the safety message. The turnaround manager should also regularly take part in formal site safety inspections and spot checks.

Permits to work

Permit to Work issuers must be visited and any problems discussed. Permits are, on many overhauls, a constant source of frustration and wasted time. Given that the permit system has been set up to cater for the unique conditions of the event, permits should be available when required. If they are not, the root causes of the delays must be exposed and dealt with. With the increasing use of contractors, the client management are often faced with large claims for delays due to late permits. There is no easy answer to this. Only real co-operation between all parties concerned, constant monitoring of the situation and a willingness to act to improve when required will keep wasted time to a minimum.

Quality

There should be regular visits to the quality team to discuss any problems. The team will normally have a monitoring role in specific areas (such as the making of joints) throughout the greater part of the event and they are normally the plant representatives who sign off the contractor's work on its completion. If time pressure is resulting in shoddy workmanship (at any level) the penalty will have to be paid during start up when such workmanship will be exposed in the most expensive way – the plant will not start up or function properly. A moment's thoughtlessness can have grave consequences. Again, if there are quality problems, the root causes must be exposed and eliminated.

Working patterns

The turnaround manager should carry out a daily check on working patterns to check that they actually match the needs of the event. Shift patterns (especially those of the night shift) must be regularly checked against the amount of work which is being achieved. The daily overtime requests from Area Engineers must be analysed and approved or rejected. A daily check should be made on productivity and manpower utilization.

Co-ordination

The plant and engineering/maintenance managers must be met regularly to:

- resolve any technical problems;
- discuss, and approve or reject, requests for extra or additional work;
- formulate strategies to keep the event on schedule;
- resolve any industrial relations problems;
- resolve any interface conflicts;
- discuss and define the consequences of any change of intent.

Control (see Figure 13.4)

A daily control meeting should be chaired, with the following agenda:

1. Safety officer's report.
2. Area engineers' reports.
3. Project managers' reports.
4. Plant managers' report.
5. Engineering/maintenance manager's report.
6. Quality team's report.
7. Quantity surveyor/cost engineer's report.
8. Turnaround manager's summing up and actions.

When the meeting is over, the turnaround manager writes his daily report and issues it. The report should contain, but not be limited to, the following:

- a brief overview of the progress and performance of the turnaround highlighting any major problems or achievements;
- a safety statement highlighting good and bad practices, brief reports on accidents, the daily and cumulative safety statistics and the following day's safety theme;
- if applicable, a statement on current costs and manpower levels.

The report should be restricted to one sheet of A4 paper and circulated as widely as possible.

The daily turnaround programme (see Figure 13.5)

As well as his own daily routine, the turnaround manager should regularize the activities of the event's other key personnel by issuing a daily turnaround programme. Unlike his own routine, which can have a certain amount of flexibility from day to day, the timings of the activities on the daily programme once set, are fixed. This is because the outcomes of many of them are fed into subsequent activities. Because of the large amount of work that has to be carried out on a turnaround, having a daily programme is essential. The more 'in control' of the routine work the team is, the more time is available for dealing with the unexpected.

Information and actions for consideration

1. Safety officer's report

- details of accidents/incidents in last 24 hours
- details of accident/incident trends
- findings of daily site inspection and spot check
- summary of site safety level and details of any particular safety concerns
- recommendations for safety improvements
- details of any safety initiatives or awards
- recommendation for tomorrow's safety theme

3. Project manager's report

- progress to date on project including any technical problems and solutions
- progress on any 'break ins'
- any 'bad fit' problems due to poor design
- any hold ups or shortages of manpower, materials, equipment or services
- any conflicts with other areas of work
- whether the area is on schedule or behind – and, if behind, the strategy for getting back on target
- assessment of unavoidable over-run, how many hours or days, and why it is unavoidable

5. Maintenance manager's report

- any concerns on turnaround progress
- any engineering concerns
- any quality performance concerns
- any upcoming engineering problems
- any questions on turnaround engineering work being done

7. Area engineers' reports

- progress on major tasks including any technical problems and solutions
- tasks completed, boxed up and handed back and percentage completion of other major tasks
- progress on small tasks and bulk work
- any hold-ups or shortages of manpower, materials, equipment or services
- any conflicts with other areas of work
- whether the area is on schedule or behind – and, if behind, the strategy for getting back on target
- assessment of unavoidable over-run, how many hours or days, and why it is unavoidable

2. Turnaround manager's routine

- open the meeting and control through the chair
- ask for and note reports in pre-set order
- ask specific questions to clarify points
- do not allow detailed discussion of issues at this meeting – convene separate discussions
- sum up general progress on key indicators
- voice any concerns on trends or specific issues
- make executive decisions and inform the meeting of them, their requirements and consequences
- delegate specific actions to particular people
- delegate responsibilities to convene further discussions on key issues outside of the control meeting
- announce next day's quality initiative
- announce next day's safety slogan
- make any other announcements
- state, and ask for, any other business
- close the meeting

4. Quality team leader's report

- quality trends in the last 24 hours
- any specific quality problems
- any recurring quality problems
- recommendations for quality improvement

6. Plant manager's report

- current ability of permit to work issuers to issue permits on time and strategy to eliminate any delays
- any hand over quality issues
- general view of on-site performance
- general view of on-site housekeeping
- upcoming on-site problems
- warning of any system coming back on line early
- warning of any process activity that could impact progress or safety

8. Quantity surveyor's report

- actual vs planned expenditure to date
- expenditure trends in each area
- specific examples of cost over-run
- general forecast on final turnaround cost
- any recommendations for tighter cost control or cost saving initiatives

Figure 13.4 The daily control meeting

Time	Activity	Personnel
0700	Report on overnight activities to area co-ordinators	night shift engineer, co-ordinators
0715	Prepare permits to work and issue Marshal materials and equipment and collect permits – review task sheets	permit issuers, co-ordinators, supervisors
0730	Brief area team on day's work	area engineer
0800	Labour reports for work and are 'checked' on the job	supervisors all trades
0805	Work teams briefed on requirements	supervisors
0815	Day's work commences	all personnel
1015	Morning coffee break	all personnel
1130	Formulate overtime requests for that evening	co-ordinators, supervisors
1145	Approve/modify overtime request	manager
1300	Lunch break (Thirty minutes)	all personnel
1430	Update of work progress against plan for all areas of the turnaround	area planners, supervisors
1500	Site logistics meeting to programme following day's crane lifts and other rigging requirements	logistics officer, crane co-ordinator, rigging foremen
1530	Area team meetings to programme last twenty-four hours progress and safety issues	area engineer, area planners
1545	Feedback on day's work to work teams	supervisors
1600	End of normal day's work. Men are 'checked off' the job	supervisors, all trades
1600	Turnaround control meeting to report to manager on progress, raise issues and take action	manager/engineers safety officer, plant management
1630	Overtime teams commence work	supervisors + men
1930	Handover briefing for nightshift team's work programme	supervisors, night shift engineer

Figure 13.5 Example of a daily turnaround programme

The unexpected (see Figure 13.6)

The purpose of planning and preparation is to reduce what is unknown to an absolute minimum, and to routinize everything that is known. As discussed in Chapter 1, there are several uncertainties that lie at the heart of a turnaround, because it involves a plant that is worn or damaged, and to an extent that is unknown. Also, it is prone to the inherent vagaries of all human activities – we disagree, make mistakes, get tired, and change our minds without warning.

Because the turnaround is a task-centred event one uncertainty which can have a significant effect is emergent work, which falls into two categories: extra and additional, which were defined in an earlier chapter. For the purposes of this section, however, extra work is defined as that which is generated by an existing task (such as repairing a crack found during an inspection) and additional work as tasks that were not part of the original plan but were inserted or requested during the turnaround, either because they were missed off of the original work list or they have been generated by a change of intent or a by a previously sound piece of equipment failing during the event.

Outline procedure for handling technical problems

1.   A fault is discovered during the execution of work or inspection

2.   The area engineer analyses the fault to define the extent and consequences of the problem

3.   Area engineer contacts turnaround, plant and maintenance managers and appropriate technical experts and convenes a meeting

4.   Area engineer presents the problem to management team and appropriate technical experts and suggests possible solutions

5.   If necessary, a turnaround policy team meeting is convened and the team informed of the problem and possible solutions

6.   The team decides upon a course of action and responsibilities are delegated as follows

7.   Technical experts produce necessary instructions, drawings and documentation needed to carry out work

8.   Site logistics officer procures material and equipment

9.   Area engineer and planning officer calculate the effect of the extra work on the turnaround schedule

10.   Turnaround manager and quantity surveyor calculate costs

11.   Area engineer gathers all necessary information and forms it into a proposal defining work requirements, costs, timing and duration of the work

12.   Management team reconvenes to approve, rework or reject the proposal

– if approved, the work is carried out on the due date

– if reworked, the proposal is then approved

– if rejected, either operations 7–11 are repeated until approval is obtained or the work is not done

13.   When the work is approved, the planning officer amends the turnaround schedule to include the new work

Figure 13.6 The unexpected

The turnaround manager, in consultation with the plant and engineering/ maintenance managers, must approve all emergent work. In the case of extra work, allowance for such a contingency may have been built into the plan and budget. This will not be so with additional work unless, in the light of experience with previous turnarounds, a lump sum has been put into the budget to cover costs. However this alone will not address any negative impact on the schedule; emergent work (especially of the extra variety) may extend the duration of the turnaround. This effect is more critical when it happens near the end of the turnaround because there is then little or no time left to recover.

Emergent work may change the critical path from one task (or area) to another and it is then vitally important to closely monitor its effect. Resources may otherwise be concentrated on a task which is not currently the critical path, while the actual critical path task is robbed of the resources it so badly needs.

If the effect of emergent work is significant, the turnaround manager must have the authority to reconvene the policy team, at any time during the event, to discuss the impact of the work and make whatever changes to the policy or objectives that this necessitates. The types of change that emergent work, if it must be done, can force upon the policy team are:

- increasing the budget to cover extra expenditure;
- declaring a longer duration;
- altering the logic of the schedule;
- eliminating other, previously planned, work from the turnaround;
- altering resource levels, shift patterns or overtime levels;
- stopping work on a less critical area of the event and switching resources.

Whatever the changes, they will have a negative impact on routine and they are bound to cause frustration among those members of the team most affected. The turnaround manager must move to minimize this and motivate his workforce. Also, the costs of emergent work must be closely monitored and kept separate from the turnaround budget.

Once the emergent work has been discovered, there must be a routine put in place to handle it as professionally as possible (see Figure 13.7). This should process the work through initial identification, registration and endorsement (or rejection). Once endorsed it is technically specified by the appropriate level of planning and then submitted for final approval. If this is given the work is funded, resourced and integrated into the work schedule before being handed over to the person responsible for carrying it out. If the work is rejected at the endorsement or the approval stage this must be recorded to prove that the work has been rejected only after due consideration, and not forgotten or rejected out of hand. Even in this there must be single point responsibility.

The power of routines

By means of the routines for planned work and for coping with unexpected work when it occurs, the turnaround manager strives to keep control of the event. Very occasionally, a situation will still arise which completely swamps the system and makes the normal routines inadequate – a serious accident that causes a walk out, a defect so large that it will keep the plant off line for weeks or months. For the most part though, effective routines will handle almost any situation.

Starting up the plant (see Figure 13.8)

A point is reached in each area of the turnaround at which most of the tasks have been completed and the turnaround manager and the plant team together agree that the area may be handed back to the latter for start up. This is a critical transition phase and if it is not properly controlled, time, money and effort can be wasted.

Although it may seem to the casual observer that the start up of the plant is simply the reverse of the shutdown, there are some significant elements

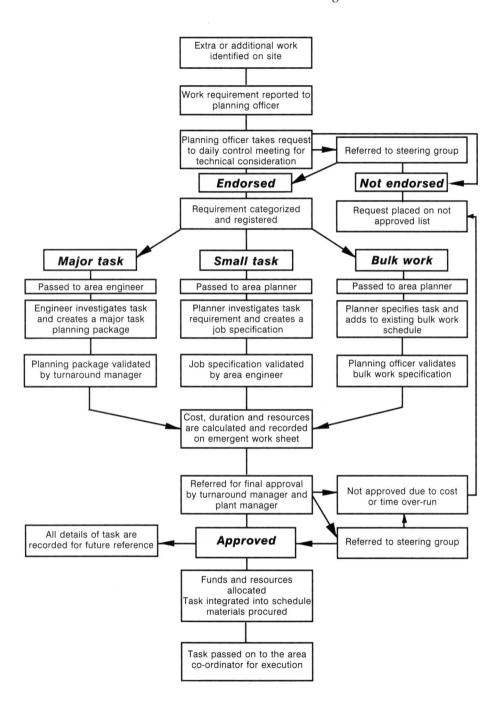

Figure 13.7 Emergent workflow diagram

- Have all items on the work list been completed?
- Has the area team carried out a quality check e.g.:

    – Does new flanged pipework have correct gasket fitted?
    – Have supports been fitted to new pipework?
    – Have new stress-relieved valves been repacked?
    – Have all joints got correct gaskets fitted?
    – Have instrument modifications been checked and tested?
    – Have all new valves been repacked/rejointed?

- What post start-up work remains, has it been listed and agreed by the start-up engineer/co-ordinator?
- Have all outstanding clearances been signed off?
- Have all control sheets been completed i.e. site mods, critical joints, vessel inspections RVs, CVs?
- Have all isolations been checked to ensure all slip plates removed?
- Do any joints in the area require 'hot flogging' – has it been arranged?
- Are the de-staging lists complete e.g. hot stagings, pre start-up staging, post start-up stagings?
- Is any special lighting required during start up?
- Has any instrument damage been identified and corrected?
- Has all equipment been removed where possible. If not, list what remains e.g. weather protection, scrap, welding sets etc?
- Do the plant operators understand the implications of the site modifications carried out?
- Can operators gain access to valves etc?
- Have the shutting down defects been identified and corrected?
- Has all extra work been executed and any start-up implications communicated?
- Are there any critical CVs/RVs which need overhauling to be available during start up?
- Are the insulation lists up to date, pre start-up/post start up?
- Are there any control valves in your area which need close attention during the start up?
- Have all electrical checks been carried out?

Figure 13.8 Handover check list: area team to start up team

which differentiate the two events, viz.:

- There are usually many tests – pressure tests, system tests, loop checks and trip and alarm tests – which have to be carried out during the start- up period which make it more complex than shutdown. Also, any one of these tests may fail and need re-working, extending the event's duration. This means that, in effect, there may be two or more work programmes running in parallel, either of which can delay the other.
- The start up comes at the end of the event when most people have been working long hours over an extended period of time. Tired people make more mistakes.
- Unlike the shutdown when the plant is being cooled and de-pressurized, and fluids are been extracted, during the start up the plant is being heated and pressurized, and fluids are being introduced, increasing the hazards.
- There is a possibility that, on shutdown, critical path time lost may

be recovered and the event put back on programme. On start up, however, such time lost cannot be recovered, it simple extends the duration.

- There have been thousands of individual activities performed on plant equipment during the shutdown and the mechanical duration. Any one of them may have been incorrectly performed and cause a fault to emerge during start up.
- An item of equipment that was not on the turnaround work list may fail on start up. This is especially true of pumps.
- An isolation plate inadvertently left in position during start up can have consequences ranging from frustrating to catastrophic.

Depending upon the type of plant, the start-up process can last anything from a few hours to many days. Whatever the length of time, the handover must be done effectively. The plant manager and his team control the start up in the same way in which they control the shutdown and, as then, the turnaround manager provides resources to support the plant team.

The turnaround manager cancels the daily control meeting and convenes in its place a daily start-up meeting, the emphasis of which is on getting the plant back on line as safely and quickly as possible. The following personnel are normally involved in the meeting:

- Plant manager   – Chairman
- Turnaround manager   – Technical adviser
- Turnaround engineer   – Work organizer
- Turnaround co-ordinator   – Punch list organizer
- Turnaround planner   – Work planner
- Logistics officer   – Material procurement
- Control instrument engineer   – Trip and alarm systems
- Safety representative   – Control of hazards
- Any other nominated person

The team meets at regular intervals during the day to review progress and update work requirements. The agenda for the meeting is as follows (see Figure 13.9):

1. Punch list of all outstanding work.
2. Additional work generated by the start up.
3. Procurement of items generated by the start up.
4. Start-up activities/key dates/problems.
5. Site clean-up.
6. Any other issues.

The start team needs to be closely co-ordinated, responding quickly to the ever-changing situation which characterizes the start up. The start up control meeting continues until the plant is back on line and productive and the plant manager is satisfied that all technical work has been completed. At that

The start-up meeting agenda in detail

1. The punch list
- specify the current list of outstanding work
- review the previous day's punch list
- strike off all completed work
- investigate causes why any planned work was not completed in the previous 24 hours
- re-define priority tasks/changes of priority
- define required completion dates/times
- highlight any hold up or shortage for each job
- highlight any job which cannot be completed
- create strategies to overcome problems
- assign resources to jobs
- responsibilities for getting work done

2. Extra/additional work
- define any extra or additional work that has been generated by the start up itself
- highlight any work which must be brought to the attention of the turnaround policy team
- add the work to the punch list
- set priorities/completion dates for the work
- if necessary, reorganize the punch list to accommodate the new work
- specify manpower, material, equipment and services requirements
- if required carry out a task hazard assessment
- (plant personnel) is there any potential extra or additional work likely to be generated in the future, based on experience of past start ups

3. Procurement
- record any deliveries in the last 24 hours
- define any existing procurement problems
- delayed/long delivery/unavailability
- delivered items rejected/need for concession or other action
- assign responsibility for procuring manpower, materials, equipment and services for newly generated work
- assign cost centres for purchase of resources

4. Start up key dates/problems
- specify current start-up logic
- list key dates for individual system start ups
- specify any change in priority from previous day's meeting
- define problems that are delaying start ups
- define any potential problem which might delay start up
- create strategies to overcome problems
- specify any unavoidable over-run of start- up programme
- if necessary, inform the turnaround policy team of expected over-run

5. Site demobilization
- demobilization of turnaround team members
- local manpower being sent off site
- contractors leaving site today
- equipment being taken off hire
- services being demobilized
- accommodation and facilities being sent off site
- return to stores/vendors all unused items and materials
- washing and cleaning programme to eliminate effluent and other unwanted substances
- housekeeping programme to return site to its original clean and safe state

6. General points
- final closing of punchlist and any exceptions
- closing out or cancellation of all work orders and permits to work
- date and time of final site inspection by plant, maintenance and turnaround managers
- date and time of post mortem debrief
- promised date for turnaround final report
- promised date for turnaround quality manual (if required)
- date/venue of turnaround topping out party

Figure 13.9 Starting the plant up

point the start-up team is de-mobilized and plant personnel return to their normal duties.

14
Terminating the turnaround

Introduction

It is the turnaround manager's responsibility to arrange all the activities that will return the plant to an acceptable condition, at least as good as it was before the event. He must ensure that the handover is properly concluded and that all traces of the event are removed, ensuring that the environment of the plant is at least as good as it was before shutdown. He must then arrange a final site inspection and obtain a handover certificate from the plant manager, indicating that the turnaround is over. Finally he must de-brief the key personnel who worked on the turnaround and, from the information gained, write a final report.

Demobilizing the site

Most of the work of demobilizing the site is arranged by the logistics officer, but it is still the responsibility of the turnaround manager to ensure that it is carried out in a proficient and proper manner. Demobilizing refers to the removal from site of all people, goods and services temporarily housed there for the duration of the event. This includes, *inter alia*:

All manpower resources, both local and contractor – The manpower demobilization should be phased to match the declining workload. The rule is to demobilize manpower at the earliest opportunity.

All surplus materials, unused proprietary items and unused consumables – These will either be returned to the stores, sold, recycled or dumped. If returned to stores, all material certificates or fitness-for-purpose certificates must be returned also. (It is ironic that some of the modern spares management systems make it difficult, if not impossible, to return surplus items to the stores.)

All tools, equipment, cranage and vehicles – Once again, hired equipment must be taken off-hire at the earliest possible date.

Turnaround control cabins and fitments – This includes cabins, furniture, computers, telephones, faxes, photocopiers etc.

All temporary accommodation – Stores, mess rooms, toilets etc.

All temporary structures – Scaffold, platforms etc. The plant team may request that some temporary   structures be left in position to allow them to perform future work. In this case the costs should be transferred to the plant budget.

Any temporary work stations – Water washing bays, quarantine compounds etc. All should be removed and thoroughly cleaned (no effluent should be left behind).

Unwanted elements – Redundant plant, spent catalyst, debris, rubbish, dirt, spillage and effluent should be safely removed.

Demobilization includes a thorough cleaning of the plant area and the removal of any other item or substance temporarily installed. In addition, any damaged painted surfaces should be patched or painted over and all insulation re-installed (except where the plant manager has requested that it be left off).

The final inspection and handover

When the turnaround manager is satisfied that the site has been fully demobilized and cleaned, he arranges a final site inspection with the plant manager, accompanied by the logistics engineer and any person nominated by the plant manager (it is also a good practice to take a small team of cleaners who can clean up any area of the plant which does not meet the plant manager's standards). They visit every area of the plant and inspect to ensure that:

- all agreed work has been completed;
- all traces of the turnaround have been removed;
- the plant is clean and tidy;
- any damage done during the turnaround has been repaired.

The plant manager should identify anything that is not satisfactory. The logistics engineer records it on a punch list. In addition, the plant manager may request certain advantageous omissions, e.g. that particular insulation be left off, or scaffolding be left in position because there is further work, unconnected with the turnaround, to be done on line. These exceptions need to be recorded because the cost of completing the work will be transferred to the plant manager's budget.

The logistics engineer arranges for any outstanding work to be carried out and, if necessary, for the turnaround manager and the plant manager to then re-inspect the plant. When the plant manager is satisfied with the state of the plant, he signs a handover certificate which indicates in a formal manner that the turnaround has been completed.

The post-mortem debrief

The turnaround has been completed. The personnel who performed it have all returned to their normal jobs. The plant is back on line. The site has been cleared and the handover certificate has been signed. However, that is not the end of the story. All of the planning and preparation that went into the event, all of the briefing and control mechanisms, were aimed at achieving one goal, performing the turnaround within the pre-set targets for safety, quality, duration and budget. The turnaround may or may not have achieved one or more of these. Some tasks will have gone well and others will have been a disappointment. Targets may have been missed for a number of reasons which only hindsight can determine. For instance:

- the targets may have been unrealistic;
- the policies imposed (by the policy team) may have been counter-productive;
- the planning and preparation may have been inadequate;
- communication may have been sub-standard;
- contractors may have been incapable of meeting their promised targets;
- emergent work may have overwhelmed the ability to meet the targets;
- unexpected events may have overwhelmed the ability to meet the targets;
- concentrating on achieving one target may have caused others to be missed.

If the organization is to be improved, it must analyse the recently completed turnaround – to measure actual against planned performance; to ascertain the root causes of success and failure so that features that caused good performance may be reinforced, and those that caused bad eliminated. This is the business of the post-mortem debrief.

A word of caution here. There is always the risk that a debrief can turn into a witch hunt. If an aspect of the event has gone badly and management are more interested in finding someone to blame for poor performance, rather than finding the root causes of the fault, the whole affair may degenerate into a blood letting. Everyone becomes defensive and nothing is learned. Heads rolling (whether the victim was actually the guilty party or not) may satisfy certain individual managers within the company, but will not promote consolidation of the wealth of information. The negative approach does the company no good and it will remain ignorant of the causes of poor performance, especially when the whole process will have to be repeated in two, three or four years' time, with the risk that the same mistakes will be made again.

It is vital that the turnaround manager sets the trend by ensuring everyone understands that the debrief, when it is held, will be a learning event, not a search for the guilty. He should agree a date for the debrief with the plant manager and any key personnel required to attend. The meeting should be chaired by him and subjects that might well be being dealt with could include:

- The scope of work on the original schedule. The amount of extra and additional work that was carried out and the reasons why it was necessary.
- Planned versus actual duration and the reasons for any differences between them.
- Planned versus actual man-hours, resources, shift patterns, overtime levels and the reasons for any differences between them.
- Targeted quality performance versus actual performance and the reasons

for any failures to meet the targets.
- The actual safety record – highlighting any accidents or incidents and the findings of the subsequent inquiries.
- General impressions on overall performance.
- Lessons to be learned from the analysis.
- Recommendations for future improvement.
- Closing comments by plant manager and turnaround manager.

NB The above list is not intended to be exhaustive.

The number of debriefing meetings required will depend upon the number of people involved.

The final turnaround report

The turnaround manager now has all the available information concerning the event. His final task is to organize it into a turnaround report. Typically, the general topics covered would be:

- Turnaround policy
- The work scope
- The preparation phase
- Planning
- The organization
- Control of work
- Contractor performance
- Safety
- Quality
- Site logistics
- Communications
- Recommendations

Figure 14.1 is a detailed checklist of the requirements for each topic.

Once the report is completed the turnaround manager issues copies to all key personnel. It is the link between one turnaround and the next, which ensures (if it is used properly) that the lessons of the past will contribute to an improved performance in the future. It is the responsibility of all key personnel to examine the report, analyse its message and recommendations and add to the sum of knowledge on which the company depends for its success and survival.

The turnaround report is the final activity in a process which has lasted many months: a process which, when driven by a rational methodology, goes a long way towards ensuring that the prior objectives of the turnaround are met to the satisfaction of the management of the company.

The final report should contain the following information

1. Turnaround policy
Breakdown of actual performance measured against initial objectives

- actual vs planned duration
- actual vs planned cost
- actual vs planned work scope including extra/ additional work
- actual vs planned man-hours including overtime and shifts
- actual vs planned accidents/incidents

2. The work scope
- a description of each major task
- a list of minor tasks
- numbers of bulk work jobs
- a description of each project
- a list of statutory inspections
- any registered inspections
- instrument and electrical work including trip and alarm tests
- a list of defects

3. The preparation phase
- turnaround management and planning hours expended
- pre-shutdown work
- long delivery items
- systems set up for reception storage and issue of materials
- a copy of the approved plot plan
- special problems encountered

4. Planning
- planning system used and any comments on its performance
- comments on number and quality of task sheets and lists
- comments on number and quality of job logic networks
- plant validation of task sheets
- any special problems

5. The organization
- a copy of the organization sheet
- comments on the size and shape of the organization
- comments on the effectiveness of the organization
- comments on work management teams
- comments on plant and specialist support

6. Control of work
- computer printed schedules
- check lists
- control sheets
- 'planning boards'
- 'S' curves
- daily meetings
- effectiveness of reporting

7. Contractor performance
List individual contractors and for each, detail actual performance against planned performance for:

- duration of work
- quantity of work done
- quality of work
- overall cost (including extras)
- safety

Score each contractor (out of 100) and recommend if they should be employed again

8. Safety
- summarize safety performance
- list each accident/incident in detail with causes and inquiry recommendations
- comment on housekeeping
- comment on wearing of PPE
- compare contractor performance with local resource performance
- comment on effectiveness of the permit to work system
- comment on the effectiveness of the emergency system

9. Quality
- define quality team function
- comment on quality team performance
- highlight any major quality non conformances
- comment on contractor quality performance
- comment on any specific quality initiatives

10. Site logistics
- outline the site logistics plan
- comment on actual facilities vs planned facilities
- comment on performance of the site logistics team
- highlight any specific difficulties

11. Communications
- describe briefing programme
- comment on turnaround team internal communication
- comment on plant/turnaround team communication
- comment on turnaround team/contractor communication
- comment on communication of daily safety initiatives
- any other general comments

12. Recommendations
Write a conclusion on the overall effectiveness of the turnaround. Recommend improvements in any of the eleven categories covered by the report – based upon:
- any safety inquiry
- any technical inquiry
- the post mortem de-brief
- personal experience
- the experience of anyone who volunteered recommendations

Figure 14.1 The final turnaround report

The next action is to decide the date for the first meeting of the policy team for the next turnaround, and the first document which should be considered at that meeting is the report from the previous turnaround.

Index